Lecture Notes in Mathematics

1578

Editors:
A. Dold, Heidelberg
B. Eckmann, Zürich
F. Takens, Groningen

Joseph Bernstein   Valery Lunts

Equivariant Sheaves and Functors

Springer-Verlag

Berlin Heidelberg New York
London Paris Tokyo
Hong Kong Barcelona
Budapest

Authors

Joseph Bernstein
Department of Mathematics
Harvard University
Cambridge, MA 02138, USA

Valery Lunts
Department of Mathematics
Indiana University
Bloomington, IN 47405, USA

Mathematics Subject Classification (1991): 57E99, 18E30

ISBN 3-540-58071-9 Springer-Verlag Berlin Heidelberg New York
ISBN 0-387-58071-9 Springer-Verlag New York Berlin Heidelberg

CIP-Data applied for

© Springer-Verlag Berlin Heidelberg 1994
Printed in Germany

SPIN: 10130027   46/3140-543210 - Printed on acid-free paper

Contents

Introduction.

Let $f : X \to Y$ be a continuous map of locally compact spaces. Let $Sh(X)$, $Sh(Y)$ denote the abelian categories of sheaves on X and Y, and $D(X)$, $D(Y)$ denote the corresponding derived categories (maybe bounded $D = D^b$ or bounded below $D = D^+$ if necessary). It is well known that there exist functors

$$f^*, \; f_*, \; f^!, \; f_!, \; D, \; Hom, \; \otimes$$

between the categories $D(X)$ and $D(Y)$, which satisfy certain identities.

Now assume that X, Y are in addition G-spaces for a topological group G, and that f is a G-map. Instead of sheaves let us consider the equivariant sheaves $Sh_G(X)$, $Sh_G(Y)$. One wants to have triangulated categories $D_G(X)$, $D_G(Y)$ – "derived categories of equivariant sheaves" – together with all the above functors. More precisely, there should exist the forgetful functor

$$For : D_G \to D,$$

so that the functors in categories D_G are compatible with the usual ones in categories D under this forgetful functor. Simple examples show that the derived category $D(Sh_G)$ of the abelian category Sh_G cannot be taken for D_G (unless the group G is discrete). The main purpose of this work is to introduce the suitable category D_G and to define the corresponding functors.

Actually, we get more structure. Namely, let $\phi : H \to G$ be a homomorphism of groups, X be an H-space, Y be a G-space, and $f : X \to Y$ be a map compatible with the homomorphism ϕ. In this situation we have functors of inverse and direct image

$$Q_f^* : D_G(Y) \to D_H(X),$$
$$Q_{f*} : D_H(X) \to D_G(Y).$$

The direct image functor Q_{f*} is probably the most interesting one. It does not in general commute with the forgetful functor.

For a connected Lie group G we give an algebraic description of the triangulated category $D_G(pt)$ in terms of DG-modules over a natural DG-algebra $\mathcal{A}_G$. This description is our main tool in applications of the theory. As an example of an application we "compute" the equivariant intersection cohomology (with compact supports) of toric varieties.

Let us explain briefly the structure of the text. Part I is devoted mainly to the definition of the category $D_G(X)$ and of various functors. In part II we use DG-modules to study the category $D_G(pt)$ and discuss equivariant cohomology. Finally, in the last part III the general theory is applied to toric varieties, which yields some applications to combinatorics.

This text summarizes the work which started some five years ago. During this period the authors were partially supported by the NSF.

Part I. Derived category $D_G(X)$ and functors.

0. Some preliminaries.

0.1. Let G be a topological group and X be a topological space. We say that X is a G-**space** if G acts continuously on X. This means that the multiplication map

$$m : G \times X \to X, \quad (g, x) \mapsto gx$$

is continuous.

Let X, Y be G-spaces. A continuous map $f : X \to Y$ is called a G-**map** if it commutes with the action of G on X and Y.

More generally, let $\phi : H \to G$ be a homomorphism of topological groups. Let X be an H-space and Y be a G-space and $f : X \to Y$ be a continuous map. We call f a ϕ-**map** if

$$f(hx) = \phi(h)f(x)$$

for all $x \in X$, $h \in H$.

Let X be a G-space. We denote by $\overline{X} := G \backslash X$ the quotient space (the space of G-orbits) of X and by $q : X \to \overline{X}$ the natural projection. By definition q is a continuous and open map.

0.2. Let X be a G-space. Consider the diagram of spaces

$$G \times G \times X \underset{\substack{\xrightarrow{d_1}\\ \xrightarrow{d_2}}}{\xrightarrow{d_0}} G \times X \underset{\substack{\xleftarrow{s_0}\\ \xrightarrow{d_1}}}{\xrightarrow{d_0}} X$$

where

$$d_0(g_1, \ldots, g_n, x) = (g_2, \ldots, g_n, g_1^{-1}x)$$

$$d_i(g_1, \ldots, g_n, x) = (g_1, \ldots, g_i g_{i+1}, \ldots, g_n, x), \quad 1 \le i \le n-1$$

$$d_n(g_1, \ldots, g_n, x) = (g_1, \ldots, g_{n-1}, x)$$

$$s_0(x) = (e, x)$$

A G-**equivariant sheaf** on X is a pair (F, θ), where $F \in Sh(X)$ and θ is an isomorphism

$$\theta : d_1^* F \simeq d_0^* F,$$

satisfying the cocycle condition

$$d_0^* \theta \circ d_2^* \theta = d_1^* \theta, \quad s_0^* \theta = id_F.$$

We will always assume that F is an abelian sheaf or, more generally, a sheaf of R - modules for some fixed ring R.

A **morphism** of equivariant sheaves is a morphism of sheaves $F \to F'$ which commutes with θ.

Equivariant sheaves form an abelian category which we denote by $Sh_G(X)$.

Examples.
   1. $Sh_G(G) \simeq R - mod$.
   2. If G is a connected group, then $Sh_G(pt) \simeq R - mod$.

Remark. In case G is a discrete group, a G-equivariant sheaf is simply a sheaf F together with an action of G which is compatible with its action on X (cf. [Groth]).

0.3. Consider the quotient map $q : X \to \overline{X}$. Let $H \in Sh(\overline{X})$. Then $q^*(H) \in Sh(X)$ is naturally a G-equivariant sheaf. This defines a functor

$$q^* : Sh(\overline{X}) \to Sh_G(X).$$

Let $F \in Sh_G(X)$. Then the direct image $q_*F \in Sh(\overline{X})$ has a natural action of G. Denote by $q_*^G F = (q_*F)^G$ the subsheaf of G-invariants of q_*F. This defines a functor

$$q_*^G : Sh_G(X) \to Sh(\overline{X}).$$

Definition. A G-space X is **free** if
   a) the stabilizer $G_x = \{g \in G | gx = x\}$ of every point $x \in X$ is trivial, and
   b) the quotient map $q : X \to \overline{X}$ is a locally trivial fibration with fibre G.

A free G-space X is sometimes called a principal G-homogeneous space over $\overline{X}$.

The following lemma is well known.

Lemma. Let X be a free G-space. Then the functor $q^* : Sh(\overline{X}) \to Sh_G(X)$ is an equivalence of categories. The inverse functor is $q_*^G : Sh_G(X) \to Sh(\overline{X})$.

0.4. The last lemma shows that in case of a free G-space we may identify the equivariant category $Sh_G(X)$ with the sheaves on the quotient $Sh(\overline{X})$. Hence in this case one may define the **derived category** $D_G(X)$ of equivariant sheaves on X to be the derived category of the abelian category $Sh_G(X)$, i. e.

$$D_G(X) := D(Sh_G(X)) = D(Sh(\overline{X})).$$

If X is not a free G-space, the category $D(Sh_G(X))$ does not make much sense in general. (However, it is still the right object in case G is a discrete group (see section 8 below)).

It turns out that in order to give a good definition of $D_G(X)$ one has first of all to resolve the G-space X, i.e. replace X by a free G-space, and then to use the above naive construction of D_G for a free space. This allows us to define all usual functors in D_G with all usual properties.

It is possible to give a more abstract definition of D_G using simplicial topological spaces (see Appendix B). However, we do not know how to define functors using this definition and hence never use it.

1. Review of sheaves and functors.

This section is a review of the usual sheaf theory on locally compact spaces and on pseudomanifolds. The subsections on the smooth base change (1.8) and on acyclic maps (1.9) will be especially important to us. We will mostly follow [Bo1].

1.1. Let X be a topological space. We fix a commutative ring R with 1 and denote by C_X the constant sheaf of rings on X with stalk R. We denote by $Sh(X)$ the abelian category of C_X-modules (i.e., sheaves of R-modules) on X.

Let $f : X \to Y$ be a continuous map of topological spaces. We denote by $f^* : Sh(Y) \to Sh(X)$ the inverse image functor and by $f_* : Sh(X) \to Sh(Y)$ the direct image functor. The functor f^* is exact and $f^*(C_Y) = C_X$. The functor f_* is left exact and we denote by $R^i f_*$ its right derived functors.

Our main object of study is the category $D^b(X)$ - the bounded derived category of $Sh(X)$. We also consider the bounded below derived category $D^+(X)$.

A continuous map $f : X \to Y$ defines functors

$$f^* : D^b(Y) \to D^b(X) \quad \text{and} \quad Rf_* : D^+(X) \to D^+(Y).$$

Remark. Since we mostly work with derived categories, we usually omit the sign of the derived functor and write f_* instead of Rf_*, $\otimes$ instead of $\overset{L}{\otimes}$ and so on.

1.2. Truncated derived categories (see [BBD])

For any integer a we denote by $D^{\leq a}(X)$ the full subcategory of objects $A \in D^+(X)$ which satisfy $H^i(A) = 0$ for $i > a$. The natural imbedding $D^{\leq a}(X) \to D^+(X)$ has a right adjoint functor $\tau_{\leq a} : D(X)^+ \to D^{\leq a}(X)$, which is called the truncation functor.

Similarly we define the subcategory $D^{\geq a}(X) \subset D^+(X)$ and the truncation functor $\tau_{\geq a} : D^+(X) \to D^{\geq a}(X)$.

Given a segment $I = [a, b] \subseteq \mathbf{Z}$ we denote by $D^I(X)$ the full subcategory $D^{\geq a}(X) \cap D^{\leq b}(X) \subset D^b(X)$.

Subcategories $D^{\geq a}(X)$, $D^{\leq b}(X)$ and $D^I(X)$ are closed under extensions (i.e. if in an exact triangle $A \to B \to C$ objects A and C lie in a subcategory, then B also lies in the subcategory). All these subcategories are preserved by inverse image functors.

If $J \subset I$, we have a natural fully faithful functor $D^J(X) \to D^I(X)$. The category $D^b(X)$ can be reconstructed from the system of finite categories $D^I(X)$, namely

$$D^b(X) = \lim_I D^I(X).$$

Since all functors $D^J(X) \to D^I(X)$ are fully faithful, there are no difficulties in defining this limit.

In the case when $I = [0,0]$ the subcategory $D^I(X)$ is naturally equivalent to $Sh(X)$. This is the heart of the category $D^b(X)$ with respect to t-structure defined by truncation functors τ (see [BBD]).

1.3. We assume that the coefficient ring R is noetherian of finite homological dimension (in fact we are mostly interested in the case when R is a field, usually of characteristic 0). Then we can define functors of tensor product $\otimes : D^b(X) \times D^b(X) \to D^b(X)$ and $Hom : D^b(X)^0 \times D^+(X) \to D^+(X)$ (see [Bo1], V.6.2 and V.7.9).

1.4. For locally compact spaces one has additional functors $f_!, f^!$ and the Verdier duality functor D. In order to define these functors we will work only with a special class of topological spaces. Namely, we say that a topological space X is **nice** if it is Hausdorff and locally homeomorphic to a pseudomanifold of dimension bounded by $d = d(X)$ (see [Bo1]).

Every nice topological space is locally compact, locally completely paracompact and has finite cohomological dimension (see [Bo1]). In particular every object in $D^b(X)$ can be realized by a bounded complex of injective sheaves. In fact we could consider instead of nice spaces the category of topological spaces satisfying these properties.

Let $f : X \to Y$ be a continuous map of nice topological spaces. Then following [Bo1] we define functors $f_*, f_! : D^b(X) \to D^b(Y)$, and $f^*, f^! : D^b(Y) \to D^b(X)$.

Functors described above are connected by some natural morphisms. We will describe some of them; one can find a pretty complete list in [GoMa]. These properties are important for us since we would like them to hold in the equivariant situation as well.

We denote by $\mathcal{T}$ the category of topological spaces.

In the rest of this section 1 (except for 1.9) we assume that all spaces are nice.

1.4.1. We have the following natural functorial isomorphisms.

$$Hom(A \otimes B, C) \simeq Hom(A, Hom(B, C)).$$

$$f^*(A \otimes B) \simeq f^*(A) \otimes f^*(B).$$

1.4.2. Composition. Given continuous maps $f : X \to Y$ and $g : Y \to Z$ there are natural isomorphisms of functors $(fg)^* = g^* \cdot f^*, (fg)^! = g^! \cdot f^!, (fg)_* = f_* \cdot g_*, (fg)_! = f_! \cdot g_!$.

1.4.3. Adjoint functors. The functor f^* is naturally left adjoint to f_* and the functor $f_!$ is naturally left adjoint to $f^!$.

1.4.4. There is a canonical morphism of functors $f_! \to f_*$ which is an isomorphism when f is proper.

1.4.5. Exact triangle of an open subset. Let $U \subset X$ be an open subset, $Y = X \setminus U, i : Y \to X$ and $j : U \to X$ natural inclusions. Then for every $F \in D^b(X)$ adjunction morphisms give exact triangles

$$i_! i^! F \to F \to j_* j^* F$$

and

$$j_! j^! F \to F \to i_* i^* F.$$

In this case $i_! = i_*$ and $j_!$ are extensions by zero, $j^* = j^!$ is the restriction to an open subset, $i_! i^!$ is the derived functor of sections with support in Y.

1.4.6. Base change. In applications we usually fix a topological space S (a base) and consider the category T/S of topological spaces over the base S. An object of this category is a pair $X \in T$ and a map $X \to S$.

Every continuous map $\nu : T \to S$ defines a base change $\sim: T/S \to T/T$ by $X \mapsto \widetilde{X} = X \times_S T$.

Given a space X/S we will use the projection $\nu : \widetilde{X} \to X$ to define a base change functor $\nu^* : D^b(X) \to D^b(\widetilde{X})$. This functor commutes with functors f^* and $f_!$, i.e. there are natural functorial isomorphisms

$$\nu^* f^* = f^* \nu^* \quad \text{and} \quad \nu^* f_! = f_! \nu^*.$$

Similarly, there are natural isomorphisms

$$\nu^! f^! = f^! \nu^! \quad \text{and} \quad \nu^! f_* = f_* \nu^!.$$

1.4.7. Properties of the functor $f^!$.

The object $D_f := f^!(C_Y) \in D^b(X)$ is called the dualizing object of f.

1. We say that the map f is locally fibered if for every point $x \in X$ there exist neighbourhoods U of x in X and V of $y = f(x)$ in Y such that the map $f : X \to Y$ is homeomorphic to a projection $F \times V \to V$.

Assume that f is locally fibered. Then for every $A \in D^b(Y)$ there is a natural isomorphism

$$f^!(A) \simeq f^*(A) \otimes f^!(C_Y)$$

(see [Ve2]).

2. Let $f : X \hookrightarrow Y$ be a closed embedding. We say that f is relatively smooth if there exists an open neighbourhood U of X in Y, such that $U = X \times \mathbf{R}^d$ and f is the embedding of the zero section $f(x) = (x, 0)$. Let $p : U \to X$ be the projection. An object $F \in D^b(Y)$ is called smooth relative to X if $F_U = p^*F'$ for some $F' \in D^b(X)$.

Assume $f : X \to Y$ is a relatively smooth embedding. Then $D_f \in D^b(X)$ is invertible (see 1.5 below). Let $F \in D^b(Y)$ be smooth relative to X. Then we have a natural isomorphism in $D^b(X)$

$$f^! F = f^* F \otimes D_f.$$

In particular the dualizing object D_Y (see 1.6.1 below) of Y is smooth relative to X and we have

$$D_X = f^! D_Y = f^* D_Y \otimes D_f.$$

3. Let

$$
\begin{array}{ccc}
Z_p & \xrightarrow{j} & Z \\
\downarrow f & & \downarrow f \\
\{p\} & \xrightarrow{i} & W
\end{array}
$$

be a pullback square, where $f : Z \to W$ is a locally trivial fibration, and $j : Z_p \to Z$ is the inclusion of the fiber. Then we have a canonical isomorphism of functors

$$j^! \cdot f^* = f^* \cdot i^!.$$

1.5. Twist. An object $L \in D^b(X)$ is called **invertible** if it is locally isomorphic to $C_X[n]$ - the constant sheaf C_X placed in degree $-n$. Then for $L^{-1} := Hom(L, C_X)$ the natural morphism $L \otimes L^{-1} \to C_X$ is an isomorphism. Every invertible object L defines a twist functor $L : D^b(X) \to D^b(X)$ by $A \longmapsto L \otimes A$. If L, M are invertible objects, then $N = L \otimes M$ is also invertible and the twist by N is isomorphic to the product of twists by L and M. In particular, the twist functor by L has an inverse given by the twist by L^{-1}.

The twist is compatible with all basic functors. For example $L \otimes (A \otimes B) \simeq (L \otimes A) \otimes B$ and $L \otimes Hom(A, B) = Hom(A, L \otimes B) = Hom(L^{-1} \otimes A, B)$.

Fix a base S and an invertible object L in $D^b(S)$. It defines a family of twist functors L in categories $D^b(X)$ for all spaces X/S; namely if $p : X \to S$ and $A \in D^b(X)$, then $L(A) = p^*(L) \otimes A$. This twist is compatible with all our functors, i.e., for every continuous map $f : X \to Y$ over the base S there are canonical isomorphisms of functors

$$f^*L = Lf^*, \quad f^!L = Lf^!, \quad f_*L = Lf_*, \quad f_!L = Lf_!.$$

These isomorphisms are compatible with isomorphisms in 1.4.

1.6. Verdier duality

1.6.1. Let us fix an invertible object D_{pt} in $D^b(pt)$ and call it a dualizing object over the point. For any nice topological space X we define its **dualizing object** $D_X \in D^b(X)$ to be $p^!(D_{pt})$, where $p : X \to pt$. If X is a smooth manifold of dimension d the dualizing object D_X is invertible (1.5) and locally isomorphic to $C_X[d]$. Using this dualizing object we define the Verdier duality functor $D : D^b(X) \to D^b(X)$ by $D(A) = Hom(A, D_X)$.

For any object $A \in D^b(X)$ we have a canonical functorial biduality morphism

$$A \to D(D(A)).$$

1.6.2. Theorem (Verdier duality). *For any continuous map f there are canonical functorial isomorphisms*

$$Df_! = f_* D \quad and \quad f^! D = Df^*.$$

1.6.3. Different choices of the object D_{pt} give rise to different duality functors, which differ by a twist. We will choose the standard normalization $D_{pt} = C_{pt}$ (see [Bo1]).

Remark. This standard normalization is not always natural. For example, if $R = k(M)$ is an algebra of functions on a nonsingular algebraic variety M, the natural choice for D_{pt} is a dualizing module for M, equal to $\Omega_M [dim M]$.

1.7. Smooth maps.

Let $f : X \to Y$ be a continuous map of topological spaces. We say that f is smooth of relative dimension d if for every point $x \in X$ there exist neighborhoods U of x in X and V of $f(x)$ in Y such that the restricted map $f : U \to V$ is homeomorphic to the projection $V \times \mathbf{R}^d \to V$.

For a smooth map f the dualizing object $D_f \in D^b(X)$ is invertible and is locally isomorphic to $C_X[d]$.

1.8. Smooth base change.

Consider a smooth base change $\nu : T \to S$. If X is a nice topological space (see 1.4.), then the space $\widetilde{X} = X \times_S T$ is also nice. The crucial observation, which makes our approach possible, is that in this situation the base change functor $\nu^* : D^b(X) \to D^b(\widetilde{X})$ essentially commutes with all other functors.

Theorem (Smooth base change).
(i) We have canonical functorial isomorphisms

$$\nu^*(A \otimes B) = \nu^*(A) \otimes \nu^*(B),$$

$$\nu^*(Hom(A,B)) = Hom(\nu^*(A), \nu^*(B)).$$

(ii) Let $f : X \to Y$ be any map of spaces over S. Let us denote by the same symbol the corresponding map $\widetilde{X} \to \widetilde{Y}$. Then for $A \in D^b(X)$, $B \in D^b(Y)$ we have canonical isomorphisms

$$\nu^* f_*(A) \simeq f_* \nu^*(A), \quad \nu^* f_!(A) \simeq f_! \nu^*(A)$$

$$\nu^* f^*(B) \simeq f^* \nu^*(B), \quad \nu^* f^!(B) \simeq f^! \nu^*(B).$$

These isomorphisms are compatible with isomorphisms in 1.4.

(iii) The Verdier duality commutes with ν^* up to a twist by the (invertible) dualizing object D_ν of $\nu : T \to S$. Namely

$$D(\nu^*(A)) = D_\nu \otimes \nu^*(D(A)).$$

This isomorphism is compatible with the identities in 1.6. For example, if we identify $\nu^*(DD(A)) \simeq DD(\nu^* A)$ using the last isomorphism then ν^* preserves the biduality morphism (1.6.1).

We will discuss this theorem in Appendix **A**.

1.9. Acyclic maps. Fix $n \geq 0$. In this section we consider general topological spaces. The proofs are given in Appendix A below.

1.9.1. Definition. We say that a continuous map $f : X \to Y$ is **n-acyclic** if it satisfies the following conditions:

a) For any sheaf $B \in Sh(Y)$ the adjunction morphism $B \to R^0 f_* f^*(B)$ is an isomorphism and $R^i f_* f^*(B) = 0$ for $i = 1, 2, \ldots, n$.

b) For any base change $\widetilde{Y} \to Y$ the induced map $f : \widetilde{X} = X \times_Y \widetilde{Y} \to \widetilde{Y}$ satisfies the property a).

We say that f is ∞-acyclic if it is n-acyclic for all n.

It is convenient to rewrite the condition a) in terms of derived categories. Namely, consider the functor $\sigma = \tau_{\leq n} \cdot f_* : D^b(X) \to D^b(Y)$. Then the adjunction morphisms $B \to f_* f^*(B)$ and $f^* f_*(A) \to A$ define functorial morphisms $\tau_{\leq n}(B) \to \sigma f^*(B)$ and $f^* \sigma(A) \to \tau_{\leq n}(A)$.

The condition a) can be now written as

a') For any sheaf $B \in Sh(Y) \subset D^b(Y)$ the natural morphism $B \to \sigma f^*(B)$ is an isomorphism.

1.9.2. It turns out that for an n-acyclic map $f : X \to Y$ large pieces of the category $D^b(Y)$ can be realized as full subcategories in $D^b(X)$. Namely, let us say that an object $A \in D^+(X)$ **comes from** Y if it is isomorphic to an object of the form $f^*(B)$

for some $B \in D^+(Y)$. We denote by $D^+(X|Y) \subset D^+(X)$ the full subcategory of objects which come from Y.

Let us fix a segment $I = [a, b] \subset \mathbf{Z}$ and consider the truncated subcategory $D^I(X|Y) = D^I(X) \cap D^+(X|Y)$.

Proposition (see Appendix **A**). *Let $f : X \to Y$ be an n-acyclic map, where $n \geq |I| = b - a$ (resp ∞-acyclic). Then*

(i) The functor $f^* : D^I(Y) \to D^I(X|Y)$ (resp. $f^* : D^+(Y) \to D^+(X|Y)$) is an equivalence of categories. The inverse functor is given by $\sigma = \tau_{\leq b} \circ f_* : D^b(X) \to D^b(Y)$ (resp. $f_* : D^+(X) \to D^+(Y)$).

(ii) The functor f^* gives a bijection of the sets of equivalence classes of exact triangles in $D^I(Y)$ and $D^I(X|Y)$ (resp. in $D^+(Y)$ and $D^+(X|Y)$). In other words a diagram (T) in $D^I(Y)$ is an exact triangle iff the diagram $f^*(T)$ in $D^I(X)$ is an exact triangle.

(iii) The subcategory $D^I(X|Y) \subset D^b(X)$ (resp. $D^+(X|Y) \subset D^+(X)$) is closed under extensions and taking direct summands.

1.9.3. The following lemma gives a criterion, when an object $A \in D^I(X)$ comes from Y.

Lemma. *Suppose we have a base change $q : \widetilde{Y} \to Y$ in which q is epimorphic and admits local sections. Set $\widetilde{X} = X \times_Y \widetilde{Y}$ and consider the induced map $\widetilde{f} : \widetilde{X} \to \widetilde{Y}$. Then*

(i) The induced map $\widetilde{f}$ is n-acyclic if and only if f is n-acyclic.

(ii) Suppose $f, \widetilde{f}$ are n-acyclic. Let $A \in D^I(X)$, where $|I| \leq n$. Then A comes from Y if and only if its base change $\widetilde{A} = q^*(A) \in D^I(\widetilde{X})$ comes from $\widetilde{Y}$.

(iii) The above assertions hold if we replace "n-acyclic" by "∞-acyclic" and D^I by D^+.

1.9.4. The following criterion, which is a version of the Vietoris-Begle theorem, gives us a tool for constructing n-acyclic maps.

We say that a topological space M is n-**acyclic**, if it is non-empty, connected, locally connected (i.e. every point has a fundamental system of connected neighborhoods) and for any R-module A we have $H^0(M, A) \simeq A$ and $H^i(M, A) = 0$ for $i = 1, 2, \ldots, n$.

Criterion. *Let $f : X \to Y$ be a locally fibered map (1.4.7). Suppose that all fibers of f are n-acyclic. Then f is n-acyclic.*

1.10. Constructible complexes.

Suppose that X is a pseudomanifold with a given stratification $\mathcal{S}$ (see [Bo1] I.1). We denote by $D_c^b(X; \mathcal{S})$ the full subcategory of $\mathcal{S}$-constructible complexes in $D^b(X)$, i.e. complexes whose cohomology sheaves are constructible with respect to S (see [Bo1]). Then $D_c^b(X; S)$ is a triangulated subcategory of $D^b(X)$, closed with respect to extensions and taking direct summands. It is also preserved by functors τ, $\otimes$, Hom and D. For a constructible complex A the biduality morphism $A \to DD(A)$ is an isomorphism. If $f : (X, S) \to (Y, S')$ is a stratified map of pseudomanifolds, then functors $f^*, f^!, f_*$ and $f_!$ preserve constructibility (see [Bo1]).

Consider the natural partial order on the set of all stratifications of X ($S \geq T$ if strata of S lie inside strata of T). If $S \geq T$ we have a natural fully faithful inclusion functor $D^b(X; T) \to D^b(X; S)$.

We define a **constructible space** to be a topological space X together with a system $\{S\}$ of stratifications of X (allowable stratifications), which is a directed system with respect to $\geq$. For constructible space X we define

$$D_c^b(X) = \lim_S D_c^b(X, S).$$

Let (X, S) and (Y, T) be constructible spaces. A continuous map $f : X \to Y$ is called constructible if for any allowable stratifications S and T there exist allowable stratifications $S' \geq S$ and $T' \geq T$ such that $f : (X, S') \to (Y, T')$ is a stratified map. For a constructible map f functors $f^*, f^!, f_*$, and $f_!$ preserve constructibility. **Examples.** 1. Let X be a complex algebraic variety. Then as a topological space X has a canonical structure of a constructible space. Namely a stratification S is allowable if all its strata are algebraic. Any algebraic map $f : X \to Y$ is constructible.

2. Similarly every real semialgebraic set has a canonical structure of a constructible space.

Appendix A.

A1. Proof of Theorem 1.8

(i) See [Bo1], V.10.1,10.21.

(ii) Consider the pullback diagram

$$
\begin{array}{ccc}
\widetilde{X} & \xrightarrow{\nu} & X \\
\downarrow f & & \downarrow f \\
\widetilde{Y} & \xrightarrow{\nu} & Y
\end{array}
$$

The isomorphism for f^* follows from 1.4.2 and for $f_!$ is a base change isomorphism 1.4.6 (here we do not use that ν is smooth).

By 1.4.7 the functor $\nu^!$ is obtained from ν^* by the twist (see 1.5) by the relative dualizing sheaf $D_\nu \in D^b(T)$, i.e., we have canonical isomorphisms $\nu^! \simeq D_\nu \otimes \nu^*$ on both $\widetilde{X}$ and $\widetilde{Y}$. Since all functors commute with the twist (see 1.5) it suffices to find canonical isomorphisms $\nu^! f^! \simeq f^! \nu^!$ and $\nu^! f_* \simeq f_* \nu^!$. Again this follows from 1.4.2 and 1.4.6.

The proof of the fact that smooth base change preserves all functorial identities mentioned in the theorem is quite lengthy and is based on case by case considerations. We omit the details.

(iii) This follows from 1.4.7 and 1.7.

A2. Proof of proposition 1.9.2

(i) Consider the functor $\sigma : D^b(X) \rightarrow D^b(Y)$ given by $\sigma(A) = \tau_{\leq b} f_*(A)$. Using adjunction morphisms $Id \rightarrow f_* f^*$ and $f^* f_* \rightarrow Id$ in combination with the truncation functor $\tau_{\leq b}$ we construct morphisms of functors $\alpha : \tau_{\leq b} \rightarrow \sigma f^*$ and $\beta : f^* \sigma \rightarrow \tau_{\leq b}$.

Let $C \subset D^b(Y)$ be the full subcategory of objects B for which the morphism α is an isomorpism. This subcategory is closed under extensions and by the acyclicity condition it contains subcategories $Sh(Y)[-i]$ for $i \geq a$. Hence C contains $D^{\geq a}$.

In particular on the category $D^I(Y)$ we have a functorial isomorphism $B \rightarrow \sigma f^*(B)$, which shows that the functor σ on this subcategory is left inverse to f^*.

Let $B \in D^I(Y)$. Properties of adjunction morphisms imply that the composition morphism $f^*(B) \rightarrow f^* \sigma f^*(B) \rightarrow f^*(B)$ is an identity. This implies the following

Criterion. An object $A \in D^I(X)$ lies in $D^I(X|Y)$ if and only if the morphism $\beta : f^* \sigma(A) \rightarrow A$ is an isomorphism.

This criterion shows that the functors f and σ are inverse equivalences of categories $D^I(Y)$ and $D^I(X|Y)$.

(ii) This follows from the fact that inverse functors f^* and σ are exact.

(iii) This follows from the criterion in (i).

The assertions about $D^+(Y)$ and $D^+(X|Y)$ in case f is ∞-acyclic are proved similarly.

A3. Proof of Lemma 1.9.3.

(i) Since $\tilde{f}$ is obtained from f by a base change, the n-acyclicity of f implies that of $\tilde{f}$ (see 1.9.1). Conversely, suppose $\tilde{f}$ is n-acyclic. Locally on Y the map $q : \tilde{Y} \to Y$ has a section s and the map f can be obtained from $\tilde{f}$ by the base change with s. Thus f is n-acyclic locally on Y and hence is n-acyclic.

(ii) If A comes from Y then clearly $\tilde{A}$ comes from Y. Let us show that if $\tilde{A}$ comes from $\tilde{Y}$ then A comes from Y.

Using the criterion in A2(i) it is enough to check this locally on Y. But locally A is obtained from $\tilde{A}$ by the base change with the morphism s, and hence it comes from Y.

(iii) Is proved similarly.

A4. Proof of criterion 1.9.4

Our proof is a refined version of the argument in [Bo1],pp.80-82.

Step 1. Let $f : X = Y \times F \to Y$ be a projection with a nonempty connected fiber F. Then for any $B \in Sh(Y)$ one has $\Gamma(X, f^*B) \simeq \Gamma(Y, B)$.

Step 2. If $f : X \to Y$ is locally fibered with locally connected fibers, then the functor f^* preserves direct products of sheaves.

Indeed, let $B = \prod B_\alpha$ be a product of sheaves on Y. Consider the natural morpism $\gamma : f^*(B) \to \prod f^*(B_\alpha)$. We claim that γ is an isomorphism. It is enough to check that γ induces isomorphisms on sections over small enough open sets $U \subset X$. By our assumption we can choose open subsets $U \subset X$ and $V \subset Y$ such that the map $f : U \to V$ is homeomorphic to a projection $V \times F \to V$ with a non-empty connected fiber F. Then by Step 1

$$\Gamma(U, f^*(B)) = \Gamma(V, B) = \prod \Gamma(V, B_\alpha)$$

and

$$\Gamma(U, \prod f^*(B_\alpha)) = \prod \Gamma(U, f^*(B_\alpha)) = \prod \Gamma(V, B_\alpha)$$

which implies that γ is an isomorphism.

Step 3. Let $i : y \to Y$ be an imbedding of a point (not necessarily closed), $j : F = f^{-1}(y) \to X$ the corresponding imbedding of the fiber over y. Then functors i_* and j_* are exact and we have the base change $f^*i_*(A) \simeq j_*f^*(A)$ for $A \in Sh(y)$.

Indeed, since the statement is local, we can assume that $X = Y \times S$. Consider a point $x = (y', s) \in X$ and a sheaf $B \in Sh(S)$. Then it is easy to see that the stalk of $j_*(B)$ at the point x equals the stalk of B at the point s if y' lies in the closure of y and equals 0 otherwise, and similarly for the map i. This implies the exactness of functors i_*, j_*. Comparison of stalks proves the base change.

Step 4. As in A2(i) let us consider the functor $\sigma = \tau_{\leq b} f_* : D(X)^b \to D^b(Y)$ and the functorial morphism $\alpha : B \to \sigma f^*(B)$.

Let $C = C(Y) \subset Sh(Y)$ be the subcategory of sheaves B such that α is an isomorphism $\alpha : B \simeq \sigma f^*(B)$.

Let $i : y \to Y$ be an imbedding of a point and $B \in Sh(y)$. Then by the assumption on fibers of the map f, the sheaf B lies in $C(y)$. Using step 3 we deduce that its image $i_*(B)$ lies in $C(Y)$. Since the functor f_* preserves direct products, step 2 implies that $C(Y)$ is closed under direct product. Hence any sheaf $B \in Sh(Y)$ can be imbedded in a sheaf E which lies in $C(Y)$.

Step 5. Using standard devissage one shows by induction on n that all sheaves on Y lie in $C(Y)$.

Step 6. For any base change $\widetilde{Y} \to Y$ the corresponding map $\widetilde{X} \to \widetilde{Y}$ satisfies the same conditions as f. Hence f is n-acyclic.

2. Equivariant derived categories.

This section contains the definition of our main object - the derived category $D_G(X)$. This is not a derived category in the usual sense, i.e. it is not the derived category of an abelian category. However, $D_G(X)$ is a triangulated category with a t-structure, whose heart is equivalent to the abelian category $Sh_G(X)$ of equivariant sheaves on X. We give several equivalent definitions of $D_G(X)$ - each one appears to be useful.

We start with the bounded derived category $D_G^b(X)$ - definition 2.2.4. Other definitions of $D_G^b(X)$ are given in 2.4 and 2.7. The bounded below category $D_G^+(X)$ is defined in 2.9 in a similar way. One notices that there is a quick definition of categories D_G^b, D_G^+ using ∞- dimensional spaces (2.7, 2.9.9). However, the most part of this section is devoted to showing that in the case of the bounded category $D_G^b(X)$ we may work only with finite dimensional spaces. This is important for the definition of functors in section 3 below. On the other hand, ∞-dimensional spaces appear to be convenient for D_G^+. In particular the definition (in section 6) of our main functor Q_{f*} - the general direct image - essentially uses ∞-dimensional spaces.

In this section we work with arbitrary topological spaces; in particular we do not assume that they are Hausdorff.

We fix a topological group G and consider the category of G-spaces.

2.1. Categories $D_G(X, P)$.

Let us recall some definitions from section 0.

2.1.1. Definition. a) A G-space is a topological space X together with a continuous (left) action of G on X. A G-map $f : X \rightarrow Y$ is a continuous G-equivariant map.

For a G-space X we denote by $\overline{X}$ the quotient space $\overline{X} = G \setminus X$ and by q the quotient map $q : X \rightarrow \overline{X}$.

b) A G-space X is called **free** if G acts freely on X and the quotient map $q : X \rightarrow \overline{X}$ is a locally trivial fibration with fiber G.

Lemma. *Let $\nu : P \rightarrow X$ be a G-map. Suppose the G-space X is free. Then $P = X \times_{\overline{X}} \overline{P}$ and in particular P is free.*

This lemma is proved in 2.3.1.

2.1.2. Definition. Let X be a G-space. A **resolution** of X is a G-map $p : P \rightarrow X$ in which the G-space P is free. A morphism of resolutions is a G-map over X.

We denote the category of resolutions of X by $Res(X)$ or $Res(X, G)$.

We will be mostly interested in resolutions which are epimorphic and moreover n-acyclic for large n.

Examples. 1. Let $T = G \times X$ be a G-space with the diagonal action of G. Then the projection $p : T \rightarrow X$ is a resolution of X, which we call the **trivial** resolution of X.

2. More generally, let M be any free G-space. Then the projection $p : X \times M \to X$ is a resolution of X.

3. If $P \to X$ and $R \to X$ are two resolutions of X, then their product $S = P \times_X R$ also is a resolution of X, which has natural projections on P and R (this is the product of P and R in the category $Res(X)$).

4. Let $f : X \to Y$ be a G-map. Then every resolution $P \to X$ can be considered as a resolution of Y. This defines a functor $Res(X) \to Res(Y)$.

This functor has a right adjoint functor $f^0 : Res(Y) \to Res(X)$. Namely for any resolution $R \to Y$ we set $f^0(R) = R \times_Y X$ (it is called the **induced** resolution of X).

2.1.3. For any resolution $p : P \to X$ of a G-space X we consider the following diagram of topological spaces

$$Q(p) : \qquad\qquad X \overset{p}{\leftarrow} P \overset{q}{\to} \overline{P} = G \backslash P.$$

Definition. We define the category $D^b_G(X, P)$ as follows:

an object F of $D^b_G(X, P)$ is a triple $(F_X, \overline{F}, \beta)$ where $F_x \in D^b(X)$, $\overline{F} \in D^b(\overline{P})$ and $\beta : p^*(F_X) \simeq q^*(\overline{F})$ is an isomorphism in $D^b(P)$.

a morphism $\alpha : F \to H$ in $D^b_G(X, P)$ is a pair $\alpha = (\alpha_X, \overline{\alpha})$, where $\alpha_X : F_X \to H_X$ and $\overline{\alpha} : \overline{F} \to \overline{H}$ satisfy $\beta \cdot p^*(\alpha_X) = q^*(\overline{\alpha}) \cdot \beta$.

Examples. If $G = \{e\}$, the category $D^b_G(X, P)$ is canonically equivalent to the category $D^b(X)$. If X is free and $P = X$, the category $D^b_G(X, P)$ is canonically equivalent to the category $D(\overline{X})$.

2.1.4. For any G-space X we define the forgetful functor $For : D^b_G(X, P) \to D^b(X)$ by $For(F) = F_X$.

2.1.5. Let $\nu : P \to R$ be a morphism of two resolutions of a G-space X. Then we define the inverse image functor $\nu^* : D^b_G(X, R) \to D^b_G(X, P)$ by $\nu^*(F_X, \overline{F}, \beta) = (F_X, \overline{\nu}^*(\overline{F}), \gamma)$, where $\overline{\nu} : \overline{P} \to \overline{R}$ is the quotient map and $\gamma = \nu^*(\beta) : p^*(F_X) = \nu^* r^*(F_X) \to \nu^* q^*(\overline{F}) = q^* \overline{\nu}^*(\overline{F})$.

2.1.6. More generally, let $f : X \to Y$ be a G-map. Suppose we are given two resolutions $p : P \to X$ and $r : R \to Y$ and a morphism $\nu : P \to R$ compatible with f (i.e., $f \cdot p = r \cdot \nu$). In this situation we define the inverse image functor $f^* : D^b_G(Y, R) \to D^b_G(X, P)$ by $f^*(F_Y, \overline{F}, \beta) = (f^*(F_Y), \overline{\nu}^*(\overline{F}), \gamma)$, where $\overline{\nu} : \overline{P} \to \overline{R}$ is the quotient map and $\gamma = \nu^*(\beta) : p^*(f^*(F_Y)) = \nu^* r^*(F_Y) \to \nu^* q^*(F) = q^* \overline{\nu}^*(\overline{F})$.

We will use this functor mostly in two situations: when $R = P$ and when $R = f^0(P)$ is the induced resolution (see 2.1.2, example 4).

2.1.7. Let $p : P \to X$ be a resolution of a G-space X and $\overline{X}$ be the quotient space of X. We define the quotient functor $q^* : D^b(\overline{X}) \to D_G^b(X, P)$ by $q^*(A) = (q^*(A), \overline{p}^*(A), \gamma)$, where $q : X \to \overline{X}$ is the quotient map, $\overline{p} : \overline{P} \to \overline{X}$ the natural projection and γ the natural isomorphism $p^* q_X^*(A) \simeq q^* \overline{p}^*(A)$.

2.2. Categories $D_G^I(X)$ and $D_G^b(X)$.

2.2.1. We want to define the equivariant derived category $D_G^b(X)$ as a limit of categories $D_G^b(X, P)$ when resolutions $P \to X$ become more and more acyclic.

We fix a segment $I = [a, b] \subset \mathbf{Z}$ and first define the category $D_G^I(X)$.

Definition. For any resolution $p : P \to X$ we define a full subcategory $D_G^I(X, P) \subset D_G^b(X, P)$ using the forgetful functor, i.e., $F \in D_G^I(X, P)$ if $F_X \in D^I(X)$ (see 1.2). For an epimorphic map p this is equivalent to the condition $\overline{F} \in D^I(\overline{P})$.

We say that a resolution $p : P \to X$ is n-acyclic if the continuous map p is n-acyclic (see 1.9). The following proposition, which we prove in 2.3.3. is central for our purposes.

Proposition. *Let $p : P \to X$ be an n-acyclic resolution, where $n \geq |I|$. Suppose that X is a free G-space. Then the quotient functor $q^* : D^I(\overline{X}) \to D_G^I(X, P)$ is an equivalence of categories (2.1.7).*

2.2.2. We will mostly work with the following corollary of the proposition, which describes how the category $D_G^I(X, P)$ depends on the resolution. Let P, R be two resolutions of a G-space X, $S = P \times_X R$ their product and $pr_R : S \to R$ the natural projection.

Corollary. *Suppose that the resolution $P \to X$ is n-acyclic, where $n \geq |I|$. Then the functor $pr_R^* : D_G^I(X, R) \to D_G^I(X, S)$ is an equivalence of categories.*

2.2.3. Fix an n-acyclic resolution $p : P \to X$, where $n \geq |I|$. For any resolution $R \to X$ we define the functor $C_{R,P} : D_G^I(X, P) \to D_G^I(X, R)$ as a composition $C_{R,P} = (pr_R^*)^{-1} \cdot pr_P^* : D_G^I(X, P) \to D_G^I(X, S) \simeq D_G^I(X, R)$, where $S = P \times_X R$ and pr_P, pr_R - projections of S on P and R. This functor is defined up to a canonical isomorphism.

Let us list some properties of this functor, which immediately follow from the definition.

(i) If $\nu : R \to R'$ is a morphism of resolutions, we have the canonical isomorphism of functors $C_{R,P} \simeq \nu^* \cdot C_{R',P}$ (2.1.5).

In particular, for any morphism $\nu : R \to P$ we have the canonical isomorphism $C_{R,P} \simeq \nu^*$.

(ii) Let Q be another n-acyclic resolution. Then we have a canonical isomorphism of functors $C_{R,P} \simeq C_{R,Q} \cdot C_{Q,P}$; since the functor $C_{P,P}$ is the identity, in this case the functor $C_{Q,P}$ is an equivalence of categories (the inverse functor is $C_{P,Q}$).

2.2.4. From now on we assume the following property of a G-space X:

$$(*) \qquad \text{For every } n > 0 \text{ there exists an } n\text{-acyclic resolution } p : P \to X.$$

Definition. For every segment $I \subset \mathbf{Z}$ we define the category $D_G^I(X)$ to be $D_G^I(X, P)$, where P is some n-acyclic resolution of X with $n \geq |I|$. As follows from 2.2.3 this category is defined up to a canonical equivalence. If $J \subset I$, we have a fully faithful functor $i : D^J(X) \to D^I(X)$, defined uniquely up to a canonical isomorphism. We define the equivariant derived category $D_G^b(X)$ to be the limit

$$D_G^b(X) = \lim_I D_G^I(X)$$

(compare with 1.2).

Passing to the limit in constructions defined in 2.1.4, 2.1.6 and 2.1.7 we define the following functors:
(i) The forgetful functor $For : D_G^b(X) \to D^b(X)$
(ii) The inverse image functor $f^* : D_G^b(Y) \to D_G^b(X)$ for a G-map $f : X \to Y$.
(iii) The quotient functor $q^* : D^b(\overline{X}) \to D_G^b(X)$.

For example, let us describe the inverse image functor. Let $F \in D_G^b(Y)$. Choose a segment $I \subset \mathbf{Z}$ such that $F \in D_G^I(Y)$. Fix $n \geq |I|$ and find an n-acyclic resolution $R \to Y$. Then by definition F is an object of $D_G^I(Y, R)$. Consider the induced resolution $P = f^0(R) \to X$ with the natural projection $\nu : P \to R$. Using the construction from 2.1.6 we define the inverse image $\nu^*(F) \in D_G^I(X, P)$. Since P is an n-acyclic resolution of X this gives an object $f^*(F) \in D_G^I(X)$.

These constructions are compatible, which means that we have canonical isomorphisms of functors $For \cdot f^* \simeq f^* \cdot For$, $q^* \cdot \overline{f}^* \simeq f^* \cdot q^*$, where $\overline{f} : \overline{X} \to \overline{Y}$ is the quotient map, and $For \cdot q^* \simeq q^*$, where $q : X \to \overline{X}$ is the quotient map.

2.2.5. Proposition. *Let X be a free G-space. Then the quotient functor $q^* : D^b(\overline{X}) \to D_G^b(X)$ is an equivalence of categories.*

Indeed, in this case X is an ∞-acyclic resolution of X, so $D_G^b(X) \simeq D_G^b(X, X) \simeq D^b(\overline{X})$.

2.3. Proofs.
2.3.1. Proof of lemma 2.1.1. Set $Y = \overline{X}$ and consider the natural map $\alpha : P \to S = X \times_Y \overline{P}$. Since G acts freely on X, α is a bijection and G acts freely on

P. In order to prove that α is a homeomorphism we can replace Y by its small open subset. Since $X \to Y$ is a locally trivial fibration with fiber G we can assume $X = G \times Y$.

Let $\pi : X \to G$ be the projection and $\mu = \pi\nu : P \to G$. Consider the continuous map $\tau : P \to P$ given by $\tau(p) = \mu(p)^{-1}(p)$. Since μ is a G-map, τ is constant on G-orbits. Hence τ induces a continuous map $\overline{\tau} : \overline{P} \to P$, such that $\tau = \overline{\tau}q$. The action of G then defines a continuous map $\beta : G \times \overline{P} \to P$. Identifying $G \times \overline{P}$ with $X \times_Y \overline{P}$ we see that β is the inverse to α. This proves lemma 2.1.1.

2.3.2. Lemma. *Let $I \subset \mathbf{Z}$ be a segment and $p : P \to X$ be an n-acyclic resolution of a G-space X, where $n \geq |I|$. Let $D^I(\overline{P}|p)$ be the full subcategory of $D^I(\overline{P})$ defined by*

$$D^I(\overline{P}|p) = \{H \in D^I(\overline{P}) | q^*(H) \in D^I(P|X)\}$$

(i.e. it consists of objects H for which $p^*(H)$ comes from X, see 1.9.2).

Then the restriction functor $D^I_G(X, P) \to D^I(\overline{P}|p)$, $F \longmapsto \overline{F}$, is an equivalence of categories. The subcategory $D^I(P|p) \subset D(\overline{P})$ is closed under extensions and taking direct summands.

Indeed, by definition an object $F \in D^I_G(X, P)$ is a triple $(F_X, \overline{F}, \beta)$ with $F_X \in D^I(X)$, $\overline{F} \in D^I(\overline{P})$ and $\beta : p^*(F_X) \simeq q^*(\overline{F})$. By Proposition 1.9.2 the functor $p^* : D^I(X) \to D^I(P|X)$ is an equivalence of categories. Hence we can describe F by a triple $(H \in D^I(P|X), \overline{F} \in D^I(\overline{P}), \beta : H \simeq q^*(\overline{F}))$.

Such a triple is determined by $\overline{F}$ up to a canonical isomorphism, so F can be described by an object $\overline{F} \in D^I(\overline{P})$ such that $q^*(\overline{F}) \in D^I(P|X)$.

Since the subcategory $D^I(P|X) \subset D^I(P)$ is closed under extensions and taking direct summands (1.9.2), the subcategory $D^I(\overline{P}|p) \subset D(\overline{P})$ also has these properties. This proves the lemma.

2.3.3. Proof of proposition 2.2.1.

By lemma 2.1.1 we have $P = X \times_{\overline{X}} \overline{P}$. Since the resolution map $p : P \to X$ is n-acyclic and $q : X \to \overline{X}$ is epimorphic and admits local sections, we can apply lemma 1.9.3. In particular, we see that the map $\overline{p} : \overline{P} \to \overline{X}$ is n-acyclic and hence $D^I(\overline{X}) \simeq D^I(\overline{P}|\overline{X}) = \{H \in D^I(\overline{P}) | H \text{ comes from } \overline{X}\}$. By Lemma 1.9.3 this last category equals $\{H \in D^I(\overline{P}) | q^*(H) \in D^I(P|X)\} \simeq D^I(\overline{P}|p)$. It remains to apply lemma 2.3.2. This proves the proposition.

2.3.4. Proof of corollary 2.2.2.

Let $F \in D^I_G(X, R)$. By definition F is described by a triple $F_X \in D^I(X)$, $\overline{F} \in D^I(\overline{R})$ and $\beta : q^*(\overline{F}) \simeq r^*(F_X)$. Applying proposition 2.2.1 to the n-acyclic resolution $p : S \to R$ of a free G-space R we can replace the object $\overline{F} \in D^I(\overline{R})$ by an object H of an equivalent category $D^I_G(R, S)$. Thus F can be described by the 5-tuple

$$(F_X, H_R, \overline{H}, \beta : r^*(F_X) \simeq H_R, \; \gamma : p^*(H_R) \simeq q^*(\overline{H})).$$

Clearly the triple (F_X, H_R, β) is determined by the object F_X up to a canonical isomorphism. Replacing in our 5-tuple this triple by F_X, we see that F can be described by a triple $(F_X, \overline{H}, \delta : s^*(F_X) \simeq q^*(\overline{H}))$, where $\delta = \gamma \cdot p^*(\beta)$. But the category of such triples is by definition $D_G^I(X, S)$. This proves the corollary.

2.4. Description of the category $D_G^b(X)$ in terms of fibered categories.

We will mostly work with another description of the equivariant derived category $D_G^b(X)$, which uses the notion of a fibered category over the category T of topological spaces. Let us recall this notion.

2.4.1. Definition. A **fibered category** C/T is a correspondence which assigns to every object $X \in T$ a category $C(X)$, and to every continuous map $f : X \to Y$ a functor $f^* : C(Y) \to C(X)$, and to every pair of composable maps $f : X \to Y$ and $h : Y \to Z$ an isomorphism $(hf)^* \simeq f^* h^*$, which satisfy the natural compatibility conditions.

We will work with the following two examples of fibered categories: $C(X) = Sh(X)$ and $C(X) = D^b(X)$.

Usually the fibered category C/T is described in a slightly different way. Namely, consider the category C defined as follows:

An object of C is a pair (X, A), $X \in T, A \in C(X)$.

A morphism $\phi : (Y, B) \to (X, A)$ in C is a pair, consisting of a continuous map $f : X \to Y$ and a morphism $\phi : f^*(B) \to A$ in $C(X)$.

A composition of morphisms is defined in a natural way.

By definition we have the natural contravariant projection functor $\pi : C \to T$. This functor completely describes the fibered category C/T, since one can reconstruct the category $C(X)$ as the fiber of π over an object X (see [Gi]).

A morphism $\phi : (Y, B) \to (X, A)$ in C is called **cartesian** if the corresponding morphism $\phi : f^*(B) \to A$ is an isomorphism. This notion can be described directly in terms of the functor $\pi : C \to T$ (see [Gi]).

Let K be any category. We call a functor $\Phi : K \to C$ **cartesian** if for any morphism $\alpha \in Mor(K)$ its image $\Phi(\alpha) \in Mor(C)$ is Cartesian.

2.4.2. Fix a fibered category $\pi : C \to T$. Let K be any category and $\Phi : K \to T$ be a covariant functor. We want to define the category $C(\Phi)$ which is the **fiber** of the fibered category C over the functor Φ.

By definition, an object $F \in C(\Phi)$ is a *cartesian* functor $F : K^0 \to C$, such that $\pi \cdot F = \Phi$. A morphism in $C(F)$ is a morphism of functors $\alpha : F \to H$ such that $\pi(\alpha)$ is the identity morphism of the functor Φ.

In other words, an object $F \in C(\Phi)$ is a correspondence which assigns to every object $a \in K$ an object $F(a) \in C(\Phi(a))$ and to every morphism $\alpha : a \to b$ in K an isomorphism $F(\alpha) : \Phi(\alpha)^* F(b) \simeq F(a)$ in $C(\Phi(a))$ such that
a) If $\alpha = id$, then $F(\alpha) = id$.
b) For any pair of morphisms $\alpha : a \to b$, $\beta : b \to c$ in K we have

$$F(\beta\alpha) = F(\alpha) \cdot \Phi(\alpha)^*(F(\beta)).$$

Examples. 1. Let K be the trivial category (one object, one morphism). Then a functor $\Phi : K \to T$ is nothing else but a topological space X and $C(\Phi) = C(X)$.
2. Let K be the category with 3 objects in which morphisms are described by the following diagram

$$\circ \xleftarrow{\;p\;} \circ \xrightarrow{\;q\;} \circ \, .$$

For any resolution $p : P \to X$ of a G-space X the diagram $Q(p)$ (2.1.3) represents a functor $\Phi : K \to T$ and the category $D^b(\Phi)$ is equivalent to $D^b_G(X, P)$.

2.4.3. Let X be a G-space. Consider the category $K = Res(X, G)$ of resolutions of X and the functor $\Phi : K \to T$, $\Phi(P) = \overline{P} = G \setminus P$.
Proposition. *The fiber $D^b(\Phi)$ of the fibered category D^b/T is naturally equivalent to the category $D^b_G(X)$.*
In other words, we can think of an object $F \in D^b_G(X) = D^b(\Phi)$ as a collection of objects $F(P) \in D^b(\overline{P})$ for all resolutions $P \to X$ together with a collection of isomorphisms $\nu^*(F(R)) \simeq F(P)$ for morphisms of resolutions $\nu : P \to R$, satisfying natural compatibility conditions.
Proof. (i) Suppose we are given an object $F \in D^b_G(X)$. For any resolution $p : P \to X$ consider the object $p^*(F) \in D^b_G(P)$. Since P is a free G-space, by proposition 2.2.5 we can find an object $F(P) \in D^b(\overline{P})$ and an isomorphism $p^*(F) \simeq q^*(F(P))$. For any morphism of resolutions $\nu : P \to R$ we have a canonical isomorphism $\nu^*(F(R)) \simeq F(P)$ in $D^b(\overline{P})$, which corresponds to a natural isomorphism $\nu^* r^*(F) \simeq p^*(F)$ in $D^b_G(P)$. This collection defines an object $F \in D^b(\Phi)$.
(ii) Conversely, let $H = \{H(P) \in D^b(\overline{P}), H(\nu)\}$ be an object of $D^b(\Phi)$. Denote by T the trivial resolution $T = G \times X \to X$. For any resolution P consider a diagram of resolutions

$$T \longleftarrow P_+ = T \times_X P \longrightarrow P,$$

in which morphisms are projections. Note that the diagram

$$\overline{T} \longleftarrow \overline{P}_+ \longrightarrow \overline{P}$$

in T coincides with the diagram $Q(p)$ in 2.1.3. Thus H defines objects $H_X = H(T) \in D^b(X)$, $\overline{H} = H(P) \in D^b(\overline{P})$ and an isomorphism $p^*(H_X) \simeq H(P_+) \simeq q^*(H)$, i.e., an object $H(P) \in D_G(X, P)$. It is clear that the collection of objects $H(P)$ is compatible with morphisms of resolutions.

Chose a segment $I \subset \mathbf{Z}$ such that $H(T) \in D^I(X)$. Choose an n-acyclic resolution $P \to X$ where $n > |I|$. Then the object $H(P) \in D_G^I(X, P)$ by definition can be considered as an object of $D_G^I(X)$.

If R is another n-acyclic resolution and $S = P \times_X R$, then the objects $pr_P^*(H(P))$ and $pr_R^*(H(R))$ in $D_G^I(X, S)$ are canonically isomorphic to $H(S)$, which shows that the objects $H(P)$ and $H(R)$ in $D_G^I(X)$ are canonically isomorphic. This defines the inverse functor $D^b(\Phi) \to D_G^b(X)$.

Remark. Let us describe explicitly the functor $D_G^b(X) \to D^b(\Phi)$.

Given an object $F \in D_G^b(X)$ we can find an n-acyclic resolution $P \to X$ with $n \geq |I|$ and realize F as an object in $D^I(\overline{P})$. For any resolution $R \to X$ consider the product resolution $S = P \times_X R$ and define an object $F(R) \in D^I(\overline{R})$ by $F(R) = (pr_R^*)^{-1} pr_P^*(F)$ (here we use the fact that the functor pr_R^* gives an equivalence of categories since the map $pr_R : \overline{S} \to \overline{R}$ is n-acyclic; see also 2.2.3). This construction gives us a collection of objects $F(R) \in D^b(\overline{R})$, i.e., an object in $D^b(\Phi)$.

2.4.4. Analyzing the proof of proposition 2.4.3 we see that in order to reconstruct the category $D_G^b(X)$ we do not need to consider all resolutions and all morphisms between them. We can work with a smaller family of resolutions and their morphisms, provided it is rich enough. The following proposition, whose proof is just a repetition of the proof in 2.4.3, gives the precise statement of the result.

Proposition. *Let J be a category and $j : J \to K = Res(X, G)$ be a functor. Consider the composed functor $\Psi = \Phi \cdot j : J \to \mathcal{T}$ and denote by j the fiber functor $j : D^b(\Phi) \to D^b(\Psi)$.*

Suppose that the pair J, j has the following properties:

a) The category J has direct products and the functor j preserves direct products.

b) The image $j(J)$ contains the trivial resolution T.

c) For every $n > 0, j(J)$ contains an n-acyclic resolution.

Then $j : D_G^b(X) \simeq D^b(\Phi) \to D^b(\Psi)$ is an equivalence of categories.

Example. Let $f : X \to Y$ be a G-map. Set $J = Res(Y)$, $j = f^0 : Res(Y) \to Res(X)$ (2.1.2 example 4), $\Psi = j \cdot \Phi$. Then $D^b(\Psi) \simeq D^b(\Phi)$. In other words in order to compute $D_G^b(X)$ it is enough to use only those resolutions which come from Y.

2.4.5. Summarizing, in order to describe an object $F \in D_G^b(X)$ it is enough to do the following:

(i) To fix some sufficiently rich category J of resolutions of X closed with respect to direct products.

(ii) For every resolution $P \in J$ to describe an object $F(P) \in D^b(\overline{P})$.

(iii) For every morphism of resolutions $\nu : P \to R$ in J to construct an isomorphism $\alpha_\nu : \nu^*(F(R)) \simeq F(P)$.

(iv) To check that the system of isomorphisms constructed in (iii) is compatible with the composition of morphisms in J. Namely, given a composition of two morphisms $P \overset{\nu}{\to} R \overset{\mu}{\to} S$, we should have an equality $\alpha_\nu \cdot \nu^* \alpha_\mu = \alpha_{\mu\nu}$.

Remark. Similar results hold for the fibered category $Sh/\mathcal{T}$. Namely, the category $Sh(\Phi)$, and more generally $Sh(\Psi)$ as in 2.4.4, is naturally equivalent to the category $Sh_G(X)$ of G-equivariant sheaves on X (0.2,2.5.3). We will prove a slightly stronger statement in Appendix **B**.

We will construct the category $D_G^+(X)$ in 2.9 by the same method.

2.4.6. Let us describe some basic functors for the equivariant derived category using the language of fibered categories.

(i) The forgetful functor $For : D_G^b(X) \in D^b(X)$ (2.1.4). Identifying $D_G^b(X)$ with the category $D^b(\Phi)$ we can describe this functor by $For(F) = F(T) \in D^b(\overline{T}) = D^b(X)$, where $T = G \times X$ is the trivial resolution of X.

(ii) The inverse image functor (2.1.6). Let $f : X \to Y$ be a G-map. It defines a functor $Res(X) \to Res(Y)$, $P \mapsto P$ (see 2.1.2, example 4). Using this functor we define the inverse image functor $f^* : D_G^b(Y) \to D_G^b(X)$ by $f^*(F)(P) = F(P) \in D^b(\overline{P})$.

We also have another description of this functor, which uses induced resolutions. Namely, consider the category $Res(Y)$ of resolutions of Y and the functor $f^0 : Res(Y) \to Res(X)$, $R \mapsto f^0(R)$ (2.1.2, example 4). For every resolution $R \to Y$ we have the canonical map $f : f^0(R) \to R$. We define the inverse image functor $f^* : D_G^b(Y) \to D_G^b(X)$ by $f^*(F)(f^0(R)) = f^*(F(R)) \in D^b\overline{(f^0(R))}$. Note that this formula defines the value of the functor $f^*(F) \in D^b(\Phi)$ not on all resolutions of X, but only on resolutions of the form $f^0(P)$. However the proposition 2.4.4 shows, that this gives a well defined object in $D_G^b(X) = D^b(\Phi)$.

(iii) The quotient functor (2.1.7). The quotient functor $q^* : D^b(\overline{X}) \in D_G^b(X)$ is defined by $q^*(A)(P) = \overline{p}^*(A) \in D^b(\overline{P})$, where $\overline{p} : \overline{P} \to \overline{X}$ is the natural map.

2.5. Truncation and the structure of a triangulated category on $D_G^b(X)$.

2.5.1. Let $F \in D_G^b(X)$. We will interpret F as a functor $P \mapsto F(P)$ on the category $Res(X)$ as in 2.4.

Fix an interval $I \subset \mathbf{Z}$ and suppose that for some resolution P for which the map $P \to X$ is epimorphic we have $F(P) \in D^I$. Then for any other resolution R,

$F(R)$ also lies in D^I, since for $S = P \times R$ the projection of $\overline{S}$ on $\overline{R}$ is epimorphic and inverse images of $F(P)$ and $F(R)$ are isomorphic.

It is clear that the full subcategory of objects $F \in D^b_G(X)$ satisfying this condition is the subcategory $D^I_G(X)$ described in 2.2.1.

Similarly we define subcategories $D^{\leq a}_G(X)$, $D^{\geq a}_G(X)$ and truncation functors $\tau_{\leq a}$ and $\tau_{\geq a}$ (for example, $\tau_{\leq a}(F)(P) = \tau_{\leq a}(F(P))$).

2.5.2. Definition. A diagram $F \to F' \to F'' \to F[1]$ in $D_G(X)$ is called an **exact triangle** if for any resolution $P \to X$ the diagram $F(P) \to F'(P) \to F''(P) \to F(P)[1]$ is an exact triangle in $D(\overline{P})$.

Proposition. *(i) The collection of exact triangles makes $D^b_G(X)$ into a triangulated category.*

(ii) The truncation functors $\tau_{\leq}$ and $\tau_{\geq}$ define a t-structure on the category $D^b_G(X)$ (see [BBD]).

Proof. (i) We have to check the axioms of triangulated categories. Each axiom deals with a finite number of objects $A_i \in D^b_G(X)$. Let us choose a segment $I \subset \mathbf{Z}$ such that all objects $A_i[k]$ for $k = -1, 0, 1$ lie in $D^I_G(X)$. Then choose $n > |I|$, fix an n-acyclic resolution $P \to X$ and identify $D^I_G(X)$ with $D^I_G(X, P)$.

As was proved in 2.3.2, the natural restriction functor $D^I_G(X, P) \to D^I(\overline{P})$ gives an equivalence of the category $D^I_G(X, P)$ with the full subcategory $D^I(\overline{P}|p) \subset D^b(\overline{P})$ which is closed under extensions.

Each axiom of triangulated categories asserts the existence of some objects and morphisms, such that certain diagrams are commutative and certain diagrams are exact triangles. We find the corresponding objects and morphisms in the category $D^I(\overline{P})$ and extend these diagrams to other resolutions of X as in remark in 2.4.3. Using proposition 1.9.2 we see that this extension preserves commutative diagrams and exact triangles.

(ii) Since all the functors in the definition of the category $D^b_G(X)$ commute with truncation functors, the assertion is obvious.

2.5.3. Proposition. *The heart $C = D^{\leq 0}_G(X) \cap D^{\geq 0}_G(X)$ of the t-category $D^b_G(X)$ is naturally equivalent to the category $Sh_G(X)$.*

We will prove this result in Appendix B.

By definition, an object $F \in D^b_G(X)$ lies in C iff $F(P)$ is a sheaf for every resolution P. In other words, we can identify the category $Sh_G(X)$ with the fiber $Sh(\Phi)$ of the fibered category Sh/T over the functor $\Phi : Res(X) \to T$ (or with the fiber $Sh(\Psi)$ as in 2.4.4).

2.5.4. There is a natural functor $i : D^b(Sh_G(X)) \to D^b_G(X)$ from the bounded derived category of equivariant sheaves to the equivariant derived category. Namely, if F is a complex of equivariant sheaves on X, then for any resolution $p : P \to X$ it defines a complex $p^*(F)$ of equivariant sheaves on P. We denote by $i(F)(P)$ the corresponding object in $D^b(\overline{P})$ (use lemma 0.3). This functor i induces an equivalence of abelian categories $i : Sh_G(X) \simeq D_G^{\leq 0}(X) \cap D_G^{\geq 0}(X)$ (2.5.3), but in general is not an equivalence of the triangulated categories. However, it is an equivalence if the group G is discrete (8.3.1).

2.6. Change of groups. The quotient and the induction equivalences.

2.6.1. Let $H \subset G$ be a subgroup and X be a G-space. Then it is intuitively clear that restricting the action of G to H one should get a restriction functor $Res_{H,G} : D_G^b(X) \to D_H^b(X)$.

Here is an explicit description of this restriction functor.

For any H-resolution $p : P \to X$ we consider the induced G-resolution $ind(P) = G \times_H P$, with the morphism $\nu : ind(P) \to X$ given by $\nu(g,l) = g(p(l))$, where $l \in P$. Note that we have a canonical isomorphism $\overline{ind(P)} = \overline{P}$.

Let $F \in D_G^b(X)$ be an object, which we interpret as a functor $P \to F(P)$ like in 2.4.3. Then we define the object $Res_{H,G}(F) \in D_H^b(X)$ by $Res_{H,G}(F)(P) = F(ind(P)) \in D^b(\overline{P})$.

In case when H is a trivial group this functor is naturally isomorphic to the forgetful functor $For : D_G^b(X) \to D_H^b(X) = D^b(X)$.

More generally, let X be an H-space and Y be a G-space. Let $f : X \to Y$ be a ϕ-map, where $\phi : H \to G$ is the embedding (0.1). In this situation we define the inverse image functor

$$f^* : D_G^b(Y) \to D_H^b(X) \quad \text{by} \quad f^*(F)(P) = (F(ind(P)).$$

(Here $ind(P) = G \times_H P$ and the G-map $\nu : ind(P) \to Y$ is given by $\nu(g,l) = gfp(l)$ for $g \in G$ and $l \in P$).

By definition the functor f^* preserves the t-structure.

2.6.2. Quotient equivalence. We saw in proposition 2.2.5. that for a free G-space X we have the equivalence $D_G^b(X) \simeq D^b(\overline{X})$. This is also clear from the description of the category $D_G^b(X)$ in terms of fibered categories, since in this case the category $Res(X,G)$ has a final object X.

Similar arguments prove the following more general result.

Let $H \subset G$ be a normal subgroup, $B = G/H$ the quotient group. Then for any G-space X the space $Z = H \backslash X$ is a B-space and the projection $q : X \to Z$ is a ϕ-map, where $\phi : G \to B$ is the quotient homomorphism.

Theorem. *If X is free as an H-space then the t-categories $D_B^b(H \backslash X)$ and $D_G^b(X)$ are naturally equivalent.*

Proof. Let $P \to X$ be a resolution of a G-space X. Then $P_0 = H \backslash P$ is a resolution of the B-space Z. This defines a functor $Res(X, G) \to Res(Z, B)$. Since X is free as an H-space, lemma 2.1.1 implies that this functor is an equivalence of categories (the inverse functor is $P_0 \mapsto P = X \times_Z P_0$). Since $\overline{P}_0 = \overline{P}$ we see that this equivalence is compatible with quotient functors $\Phi_X : Res(X) \to \mathcal{T}$ and $\Phi_Z : Res(Z) \to \mathcal{T}$. This implies that the fiber categories $D^b(\Phi_Z)$ and $D^b(\Phi_X)$ are equivalent.

2.6.3. Induction equivalence. Let $\phi : H \to G$ be a embedding of a subgroup. For an H-space X consider the induced G-space $ind(X) = G \times_H X$. We have a canonical ϕ-map $\nu : X \to ind(X), x \mapsto (e, x)$. It induces an inverse image functor $\nu^* : D^b_G(ind(X)) \to D^b_H(X)$ (2.6.1).

Theorem. *The functor ν^* is an equivalence of t-categories.*
Proof. The functor $P \to ind(P)$ is an equivalence of categories $ind : Res(X, H) \to Res(ind(X), G)$ such that $\overline{P} = \overline{ind(P)}$. Hence by definition of the functor ν^* in 2.6.1 it is an equivalence of categories.

Remark. Let X be a G-space and $H \subset G$ be a subgroup. Then X can be considered as an H-space and we have a canonical G-map $\pi : ind(X) \to X$. Using the inverse image functor and the induction equivalence we get a functor

$$\pi^* : D^b_G(X) \to D^b_G(ind(X)) \simeq D^b_H(X).$$

It is easy to see that this functor is canonically isomorphic to the restriction functor $Res_{G,H}$ (2.6.1).

2.7. Other descriptions of the category $D^b_G(X)$.
    We are going to give two more alternative descriptions of the category $D^b_G(X)$.

2.7.1. Fix a sequence of resolutions $P_0 \to P_1 \to \cdots \to P_n \to \cdots$ of X where P_n is an n-acyclic resolution (if G is a Lie group, we can take $P_i = X \times M_i$, where $\{M_i\}$ is the sequence of free G-manifolds constructed using the Stiefel manifolds as in section 3.1. below). Then we can define $D^b_G(X)$ as the 2-limit of the categories $D^b_G(X, P_i)$. In other words, an object $F \in D^b_G(X)$ is a sequence $F = \{F_X \in D^b(X), \overline{F}_n \in D^b(\overline{P}_n)\}$, together with a system of isomorphisms $p^*_n(F_X) \simeq q^*_n(\overline{F}_n)$, $\nu^*_{in}(\overline{F}_n) \simeq \overline{F}_i$, where $p_n : P_n \to X$, $q_n : P_n \to \overline{P}_n$ and $\nu_{i,n} : P_i \to P_n$, satisfying obvious compatibility conditions. Corollary 2.2.2 implies that this category is equivalent to the category $D^b_G(X)$.

2.7.2. The following description of the category $D^b_G(X)$ probably provides the most satisfactory intuitive picture.

Let us fix an ∞-acyclic locally connected free G-space M (for example, take

$$M = \varinjlim M_n$$

as in section 3.1. below). Then $P = X \times M$ is an ∞-acyclic resolution of X and by definition $D_G^b(X) = D_G^b(X, P)$.

Note that if M is in addition contractible, then the fibration $\overline{P} \to \overline{M}$ is nothing else but the standard fibration $X_G \to BG$ over the classifying space of G with the fiber X. By definition an object $F \in D_G^b(X, P)$ is essentially an object in $D^b(\overline{P})$ whose restrictions to all fibers are isomorphic.

For example, let us describe the category $D_G^b(pt)$.

Proposition. *Let M be a contractible locally connected free G-space, $BG := G \backslash M$ – the classifying space for G. Then the category $D_G^b(pt)$ is equivalent to the full subcategory of $D^b(BG)$ which consists of complexes with locally constant cohomology sheaves. If G is a connected Lie group, then this subcategory consists of complexes with constant cohomology sheaves.*

Proof. By the criterion 1.9.4 the map $p : M \to pt$ is ∞-acyclic. Hence by 2.3.2 $D_G^b(pt)$ is equivalent to the full subcategory $D^b(BG|p) \subset D^b(BG)$. We claim that an object $F \in D^b(BG)$ lies in this subcategory iff its cohomology sheaves are locally constant. Indeed, consider the object $H = q^*(F) \in D^b(M)$, where $q : M \to BG$ is the quotient map. If $F \in D^b(BG|p)$, then H comes from pt and hence its cohomology sheaves are constant. Therefore the cohomology sheaves of F are locally constant.

Conversely, suppose that F has locally constant cohomology sheaves. Let us show that $F \in D^b(BG|p)$. Since by 2.3.2 this subcategory is closed under extensions we can assume that F is a sheaf. Then H is a locally constant sheaf on a contractible space M. Hence it is constant, i.e. comes from pt.

If G is a connected Lie group, then BG is simply connected and hence every locally constant sheaf on BG is constant. This proves the proposition.

This picture will be used for example to describe the behavior of the equivariant derived category when we change the group. Unfortunately the space M in this case is usually infinite-dimensional, so it is difficult to define functors like $f^!$ or the Verdier duality using this picture. However we will use ∞-acyclic resolutions in section 2.9 below in the discussion of the category $D_G^+(X)$.

2.8. Constructible objects.

Suppose that a G-space X is a stratified pseudomanifold with a given stratification $\mathcal{S}$.

Definition. We say that an object $F \in D_G^b(X)$ is $\mathcal{S}$-constructible if the object $F_X \in D^b(X)$ is $\mathcal{S}$-constructible. We denote the full subcategory of $\mathcal{S}$-constructible objects in $D_G^b(X)$ by $D_{G,c}^b(X)$.

Remark. We do not assume that the stratification S is G-invariant.

2.9. Category $D_G^+(X)$.

Let G be a topological group with the following property:

$(**)$ There exists a $\infty-$acyclic free $G-$space M.

For example, any Lie group satisfies this condition. It follows that every G-space X has an ∞-acyclic resolution $X \times M \to X$ (1.9.4).

We now proceed exactly as in section 2.4 replacing everywhere D^b (or D^I) by D^+ and n-acyclic resolutions by ∞-acyclic ones. Namely, fix a G-space X. Consider the category $K = Res(X, G)$ of resolutions of X and the functor $\Phi : K \to T$, $\Phi(P) = \overline{P} = G\backslash P$.

2.9.1. Definition. Define $D_G^+(X) := D^+(\Phi)$ - the fiber of the fibered category D^+/T over the functor Φ (2.4.1, 2.4.2).

In other words to define an object $F \in D_G^+(X)$ means for every resolution $P \to X$ to give an element $F(P) \in D^+(\overline{P})$ and for every morphism of resolutions $\nu : P \to R$ to give an isomorphism $\nu^* F(R) \simeq F(P)$ satisfying natural compatibility conditions (2.4.2).

2.9.2. Given a resolution $P \to X$ we define the category $D_G^+(X, P)$ replacing D^b by D^+ in the definition 2.1.3.
Lemma. (A D^+-version of lemma 2.3.2.) *Let* $p : P \to X$ *be an* ∞-*acyclic resolution. Let* $D^+(\overline{P}|p) \subset D^+(\overline{P})$ *be the full subcategory consisting of objects* H *such that* $q^* H$ *comes from* X. *Then the restriction functor* $D_G^+(X, P) \to D^+(\overline{P}|p)$ *is an equivalence of categories.*

2.9.3. Let $\overline{X}$ be the quotient space. We have the obvious quotient functor q^* : $D^+(\overline{X}) \to D_G^+(X, P)$ as in 2.1.7.

The following D^+-analogues of proposition 2.2.1 and corollary 2.2.2 are proved similarly using proposition 1.9.2, lemma 1.9.3 and lemma 2.9.2 above.

Proposition. *Let* $P \to X$ *be an* ∞-*acyclic resolution. Suppose that* X *is a free* G-*space. Then the quotient functor* $q^* : D^+(\overline{X}) \to D_G^+(X, P)$ *is an equivalence of categories.*

Corollary. *Let* $P \to X$ *and* $R \to X$ *be two resolutions and* $S = P \times_X R$ *be their product. Assume that* P *is* ∞-*acyclic. Then the functor* $pr_R^* : D_G^+(X, R) \to D_G^+(X, S)$ *is an equivalence of categories.*

2.9.4. The following proposition provides a "realization" of the categoty $D_G^+(X)$ as in 2.7.2.

Proposition. *Let $P \to X$ be an ∞-acyclic resolution. Then the categories $D_G^+(X)$ and $D_G^+(X,P)$ are naturally equivalent.*

The proof is similar to the proof of the proposition 2.4.3 above.

2.9.5. Combining the above proposition with lemma 2.9.2 we get the following geometric realization of the category $DG^+(pt)$.

Proposition. *Let M be a contractible locally connected free G-space, $BG := G \backslash M$ - the classifying space for G. Then the category $D_G^+(pt)$ is naturally equivalent to the full subcategory of $D^+(BG)$ which consists of complexes with locally constant cohomology sheaves. If G is a connected Lie group, then this subcategory consists of complexes with constant cohomology sheaves.*

The proof is the same as in proposition 2.7.2.

2.9.6. The proposition 2.4.4 remains true if we replace D^b by D^+ and require that $j(J)$ contains an ∞-acyclic resolution. For example, if $P \to X$ is ∞-acyclic and $T = G \times X \to X$ is the trivial resolution, then the following subcategory of $Res(X,G)$ is rich enough to define $D_G^+(X)$:

$$T \longleftarrow T \times_X P \longrightarrow P$$

(cf. example 2.4.2 and the proof of proposition 2.4.3).

Also the discussion in 2.4.5 is valid for D^+.

2.9.7. We define exact triangles and truncation functors $\tau_{\leq}$, $\tau_{\geq}$ in $D_G^+(X)$ exactly as in 2.5.1, 2.5.2. The same proof (using ∞-acyclic resolutions) shows that D_G^+ is a triangulated category with a t-structure given by the functors τ.

2.9.8. All constructions and results of section 2.6 are valid for the category D_G^+. Namely, just replace the symbol D_G^b by D_G^+ everywhere.

2.9.9. Remark. We see that using ∞-acyclic resolutions $P \to X$ one gets a quick and "geometric" definition of the category $D_G^+(X) \simeq D_G^+(X,P) \subset D^+(\overline{P})$ (2.9.4, 2.9.2). We could do the same with the bounded category $D_G^b(X)$. However, as was already mentioned, the space P is usually ∞-dimensional, which makes it difficult to apply this construction in the algebraic situation or in the case of functors like $f^!$ or the Verdier duality.

The category D_G^+ is needed in order to define the general direct image functor Q_* (when the group changes) (section 6). Although we did not use ∞-acyclic resolutions in the definition of D_G^+ (2.9.1), they become essential in the definition of the direct image Q_*. This is unpleasant in the algebraic setting. We manage to avoid this problem in some special cases (see section 9 below).

Appendix B. A Simplicial description of the category $D_G(X)$.

B1. Let us describe the category $D_G^b(X)$ in the simplicial language. For a discussion of simplicial spaces and simplicial sheaves we refer to [D1].

Let X be a G-space. Following [D1], we denote by $[G \backslash X].$ the usual simplicial space

$$[G \backslash X]. \ = \ ...G \times G \times X \ \substack{\longrightarrow \\[-0.6em] \longrightarrow \\[-0.6em] \longrightarrow} \ G \times X \ \substack{\longrightarrow \\[-0.6em] \longleftarrow \\[-0.6em] \longrightarrow} \ X.$$

Recall that a **simplicial sheaf** $F^{\cdot}$ on $[G \backslash X].$ is a collection of sheaves $F^{\cdot} = \{F^n \in Sh(G^n \times X)\}_{n \geq 0}$ with the following additional structure. Let $h : G^n \times X \to G^m \times X$ be a composition of arrows in $[G \backslash X].$. Then h defines a structure morphism $\alpha_h : h^* F^m \to F^n$, such that

$$\alpha_{h'h} = \alpha_h \cdot h^* \alpha_{h'},$$

whenever the composition $h'h$ makes sense.

The abelian category of simplicial sheaves on $[G \backslash X].$ is denoted by $Sh([G \backslash X].)$.

B2. Definition. Denote by $Sh_{eq}([G \backslash X].)$ the full subcategory of $Sh([G \backslash X].)$ consisting of simplicial sheaves $F^{\cdot}$ for which all structure morphisms are isomorphisms.

B3. Fact. The category $Sh_{eq}([G \backslash X].)$ is naturally equivalent to the category $Sh_G(X)$ of G-equivariant sheaves on X (see [D1](6.1.2,b)).

In proposition **B4** below we extend this equivalence to derived categories.

Let $D^b([G \backslash X].)$ be the bounded derived category of simplicial sheaves on $[G \backslash X].$. Denote by $D_{eq}^b([G \backslash X].)$ the full subcategory of $D^b([G \backslash X].)$ consisting of complexes A, such that $H^{\cdot}(A) \in Sh_{eq}([G \backslash X].)$.

B4. Proposition. *Triangulated categories* $D_{eq}^b([G \backslash X].)$ *and* $D_G^b(X)$ *are naturally equivalent.*

Proof. Let us first prove a special case.

B5. Lemma. *The proposition holds if X is a free G-space.*

Proof of the lemma. Consider the quotient map $q : X \to \overline{X}$. Then q defines an augmentation of the simplicial space $[G \backslash X].$ and hence defines two functors

$$(1) \qquad q^* : Sh(\overline{X}) \to Sh_{eq}([G \backslash X].),$$

$$(2) \qquad q^* : D^b(\overline{X}) \to D_{eq}^b([G \backslash X].).$$

We know that $D_G^b(X) \simeq D^b(\overline{X})$ (2.2.5). Hence it suffices to prove that the second functor is an equivalence.

It is known (see [D1]) that the first functor is an equivalence. So it suffices to show that for any two sheaves $A, B \in Sh(\overline{X})$ we have

$$\mathrm{Ext}^i_{D(\overline{X})}(A, B) = \mathrm{Ext}^i_{D([G \backslash X].)}(q^* A, q^* B).$$

Using the standard method (see [H],7.1) we reduce the proof to the case of an elementary sheaf $A = C_U$ for an open set $U \subset S$. Then we may assume that $\overline{X} = U$ and the above equality becomes

$$H^i(\overline{X}, B) = H^i([G\backslash X]., q^*B),$$

which is well known (see[D1](6.1.2,c)). This proves the lemma.

Let X be any G-space. Let $P \to X$ be an ∞-acyclic resolution. It defines a map of simplicial spaces $p : [G\backslash X]. \to [G\backslash P].$ and hence induces the functor

$$p^* : D^b_{eq}([G\backslash X].) \to D^b_{eq}([G\backslash P].).$$

By the lemma this last category is equivalent to $D^b(\overline{P})$. So we get the functor

$$q^{*-1} \cdot p^* : D^b_{eq}([G\backslash X].) \to D^b(\overline{P}).$$

It is clear that for $F \in D^b_{eq}([G\backslash X].)$ its image $q^{*-1}p^*F$ lies in the subcategory $D^b(\overline{P}|p) \simeq D^b_G(P|p) \simeq D^b_G(X)$ (lemmas 2.3.2, 2.9.2, proposition 2.9.4). So we actually have the functor

$$(*) \qquad q^{*-1} \cdot p^* : D^b_{eq}([G\backslash X].) \to D^b_G(X).$$

We claim that it is an equivalence.

Indeed, by a simplicial version of section 1.9 we conclude that the map $p : [G\backslash P]. \to [G\backslash X].$ is ∞-acyclic, that is the functor $p^* : D^b([G\backslash X].) \to D^b([G\backslash P].)$ is fully faithful and its right inverse is p_*. This implies that the functor $(*)$ is fully faithful. On the other hand, it is clear that if $F \in Sh(X)$ is such that $p^*F \in Sh_G(P)$, then also canonically $F \in Sh_G(X)$ (use again the acyclicity of the map p). Therefore the functor $(*)$ induces the equivalence of abelian categories $Sh_G(X)$ and $D^{\leq 0}_G(X) \cap D^{\geq}_G(X)$ and hence is an equivalence. This proves the proposition.

B6. Remark. Note that the last argument also proves the proposition 2.5.3

3. Functors.

In this section we consider a G-map $f : X \to Y$ and describe functors $\otimes, Hom$, $f^*, f^!, f_*, f_!$ and D in the categories $D_G^b(X)$ and $D_G^b(Y)$. We also study relations between these functors and the ones introduced earlier in 2.6. In section 3.7 we define the integration functors $D_H^b(X) \to D_G^b(X)$ which are (left and right) adjoint to the restriction functor $D_G^b(X) \to D_H^b(X)$ for a closed subgroup $H \subset G$.

3.1. In this section 3 we assume that G is a Lie group, satisfying the following condition

(*+) For every n there exists an n-acyclic free G-space M which is a manifold.

It follows that every G-space X has an n-acyclic *smooth* resolution $M \times X \to X$ (see 1.7, 1.9, 2.1.2).

Let us show that this property holds in most interesting cases.

Lemma. *Let G be a Lie group with one of the following properties:*
  a) G is a linear group, i.e., a closed subgroup of $GL(k, \mathbf{R})$ for some k,
  b) G has a finite number of connected components.
  Then the property (+) holds for G.*
Proof. a) The Stiefel manifold M_n of k-frames in $\mathbf{R}^{n+k}$ is an n-acyclic free G-manifold.
b) By a result of G.Mostow there exists a compact Lie subgroup $K \subset G$ such that the manifold G/K is contractible. By Peter-Weyl theorem K is a linear group, so by a) it has an n-acyclic free K-manifold M'. Then $M = G \times_K M'$ is an n-acyclic free G-manifold. This proves the lemma.

For a G-space X we denote by $SRes(X)$ the category of smooth resolutions of X and smooth morphisms between them. It follows from the property (*+) that this category is sufficiently rich, so we can define $D_G^b(X)$ to be the fiber of $\pi : D^b \to T$ over the functor $\Phi_X : SRes(X) \to T, P \to \overline{P}$ (see 2.4.5). This is the definition of the category $D_G^b(X)$ which we use in order to define all functors. The main reason for sticking to smooth resolutions is the fact that all functors commute with a smooth base change (see 1.8).

3.2. Functors $\otimes$ and Hom.

Let $F, H \in D_G^b(X)$. We will consider F and H as functors on the category $SRes(X)$ of smooth resolutions of X (see 2.4.5).

Now define objects $F \otimes H$ and $Hom(F, H)$ in $D_G^b(X)$ by

$$F \otimes H(P) := F(P) \otimes H(P),$$

$$Hom(F, H)(P) := Hom(F(P), H(P))$$

for every smooth resolution P. For any smooth morphism of smooth resolutions $\nu : S \to P$ we define isomorphisms

$$F(S) \otimes H(S) \simeq \nu^*(F(P) \otimes H(P)),$$

$$Hom(F(S), H(S)) \simeq \nu^*(Hom(F(P), H(P)))$$

using the smooth base change (see 1.8).

3.3. Functors $f^*, f^!, f_*$ and $f_!$.

Let $f : X \to Y$ be a G-map of nice topological spaces. Then $P \longmapsto f^0(P)$ defines the functor $f^0 : SRes(Y) \to SRes(X)$ (see 2.1.2, example 4), which takes n-acyclic resolutions into n-acyclic ones. Hence the category $f^0(SRes(Y))$ is rich enough to define objects in $D_G^b(X)$.

We will define the above functors using the smooth base change. For example, consider the functor f_*.

Given an object $F \in D_G^b(X)$ we define $f_*(F) \in D_G^b(Y)$ by

$$f_*(F)(P) := f_*(F(f^0(P))),$$

for $P \in SRes(Y)$, where $f : f^0(P) \to P$ is the natural projection.

Given a smooth morphism $\nu : S \to P$ in $SRes(Y)$ we define the isomorphism $f_*(F)(\nu) : \nu^*(f_*(F)(P)) \simeq f_*(F)(S)$ as $f_*(F)(\nu) := f_*(F(\nu))$ using the smooth base change applied to the pullback diagram

$$
\begin{array}{ccc}
\overline{f^0(S)} & \overset{\nu}{\longrightarrow} & \overline{f^0(P)} \\
\downarrow f & & \downarrow f \\
\overline{S} & \overset{\nu}{\longrightarrow} & \overline{P}
\end{array}
$$

The collection of objects $f_*(F)(P)$ and isomorphisms $f_*(F)(\nu)$ defines an object $f_*(F) \in D_G^b(Y)$.

Similarly we define functors $f^*, f^!, f_!$. Note that functors f^* and $f_!$ can be defined using arbitrary (not necessarily smooth) resolutions.

3.4. Properties of functors $\otimes, Hom, f^*, f^!, f_*, f_!$.

3.4.1. Theorem. *(i) Let $H \subset G$ be a subgroup of G. Then the above functors commute with the restriction functor $Res_{H,G}$ (2.6.1). This means that there exist canonical isomorphisms of functors $Res_{H,G} \cdot f_* \simeq f_* \cdot Res_{H,G}$, etc. In particular, these functors commute with the forgetful functor $For : D_G^b(X) \to D^b(X)$.*
(ii) Let $H \subset G$ be a normal subgroup, $B = G/H$. Let $f : X \to Y$ be a G-map of G-spaces which are free as H-spaces. Then the above functors commute with the

quotient equivalences $q^* : D_B^b(H \setminus X) \to D_G^b(X)$ *and* $q^* : D_B^b(H \setminus Y) \to D_G^b(Y)$ *in theorem 2.6.2. In particular, when* $H = G$ *these functors commute with the quotient equivalences* $q^* : D^b(\overline{X}) \to D_G^b(X)$ *and* $q^* : D^b(\overline{Y}) \to D_G^b(Y)$.

(iii) Let $H \subset G$ *be a subgroup. Then the above functors commute with the induction functor* $D_H^b(Y) \simeq D_G^b(G \times_H Y)$ *for an* H*-space* Y *as in theorem 2.6.3.*

Proof. Immediately follows from definitions.

3.4.2. Examples.

1. We want to show that the constant sheaf C_X and the dualizing object D_X for a G-space X always have a natural structure of a G-equivariant object. Let $p : X \to pt$ be a map to a point.

We define the equivariant constant sheaf $C_X = C_{X,G} \in D_G^b(X)$ by $C_X(P) := C_{\overline{P}}$. Clearly $C_{X,G} = p^*(C_{pt,G})$ and $For(C_{X,G}) = C_X$.

Define the equivariant dualizing object on X to be $D_{X,G} = p^!(C_{pt,G}) \in D_G^b(X)$. By theorem 3.4.1 we have $For(D_{X,G}) = D_X$.

More generally, for a G-map $f : X \to Y$ we define the equivariant relative dualizing object $D_{f,G} := f^!(C_Y) \in D_G^b(X)$. Again by theorem 3.4.1 it corresponds to the usual relative dualizing object under the forgetful functor.

2. Let $Y \subset X$ be a closed G-subspace, $U = X \setminus Y$. Consider the natural imbeddings $i : Y \to X$ and $j : U \to X$. Then for $F \in D_G^b(X)$ we have the exact triangles

$$i_! i^!(F) \to F \to j_* j^*(F) \quad \text{and} \quad j_! j^!(F) \to F \to i_* i^*(F).$$

These triangles are functorial in F. They are compatible with the forgetful functor, the restriction functor $Res_{H,G}$ (2.6.1) and with the quotient and the induction equivalences of theorems 2.6.2 and 2.6.3.

3.4.3. Theorem. *All properties of the functors* $\otimes$, Hom, f^*, f_*, $f^!$, $f_!$ *listed in 1.4 hold in the equivariant case.*

Proof. Since the functorial identities listed in 1.4 commute with the smooth base change (theorem 1.8), they automatically lift to the equivariant category (see the argument in 3.3).

3.5. The equivariant Verdier duality.

Assume that X is a nice G-space (1.4).

3.5.1. Definition. Consider the equivariant dualizing object $D_{X,G} = p^! C_{pt,G} \in D_G^b(X)$ (3.4.2 example 1). For $F \in D_G^b(X)$ define its Verdier dual as

$$D(F) = Hom(F, D_{X,G}).$$

3.5.2. Theorem (The equivariant Verdier duality).

(i) There exists a canonical biduality morphism

$$F \to D(D(F))$$

in $D_G^b(X)$.

(ii) For a G-map $f : X \to Y$ we have canonical isomorphisms of functors between the categories $D_G^b(X)$ and $D_G^b(Y)$

$$Df_! \simeq f_* D \quad \text{and} \quad f^! D \simeq Df^*.$$

(iii) The Verdier duality commutes with the forgetul functor For : $D_G^b(X) \to D^b(X)$.

Proof. As in section 3.3 we work with the category $SRes(X, G)$ of smooth resolutions of X.

(i). Let us first of all analize the equivariant dualizing object $D_{X,G}$.

Let $P \to pt$ be a smooth resolution of the point pt, i. e. a smooth free G-space. Consider the induced resolution of X:

$$p^0(P) = P \times X \to X$$

and the corresponding map of quotient spaces

$$\overline{p} : \overline{p^0(P)} \to \overline{P}.$$

By definition $D_{X,G}(p^0(P)) = \overline{p}^! C_{\overline{P}} \in D^b(\overline{p^0(P)})$. Since P is a smooth free G-space, the quotient space $\overline{P}$ is also smooth. Hence the dualizing object $D_{\overline{P}}$ is invertible (1.6.1). Since $\overline{p}$ is a locally trivial fibration, by 1.4.7 we have

$$D_{X,G}(p^0(P)) := p^! C_{\overline{P}} = p^! D_{\overline{P}} \otimes p^* D_{\overline{P}}^{*-1} = D_{\overline{p^0(P)}} \otimes p^* D_{\overline{P}}^{-1}.$$

Therefore, for $F \in D_G^b(X)$, the object $D(F)(P)$ is canonically isomorphic to the usual Verdier dual $D(F(P))$ of $F(P) \in D^b(\overline{P})$ twisted by $D_{\overline{P}}^{-1}$ (1.5). In particular, $D(D(F))(P)$ is canonically isomorphic to $D(D(F(P)))$. So the usual biduality morphism

$$F(P) \to D(D(F(P)))$$

(1.6.1) induces the desired biduality morphism

$$F(P) \to D(D(F))(P).$$

It remains to check that this morphism is compatible with the smooth base change by smooth maps $R \to P$ of smooth resolutions of pt. This follows from theorem 1.8(iii). This proves (i).

(ii) is proved similarly.

(iii) follows immediately from theorem 3.4.1(i).

The theorem is proved.

Some further properties of the Verdier duality (in particular its behavior under the quotient and the induction equivalences) will be studied in section 7 below.

3.6. Equivariant constructible sheaves.

Let X be a G-space, which is a pseudomanifold with a stratification $\mathcal{S}$. Let $D_{G,c}^b(X) \subset D_G^b(X)$ be the full subcategory of G-equivariant $\mathcal{S}$-constructible objects (2.8). Then it is preserved by functors $\otimes, Hom$ and D and the biduality morphism $F \to D(D(F))$ is an isomorphism for $F \in D_{G,c}^b(X)$.

If $f : X \to Y$ is a stratified G-map of pseudomanifolds, functors $f^*, f^!, f_*$ and $f_!$ preserve constructibility. This follows from the corresponding properties of the category $D_c^b(X)$ (see 1.10), the definition of $D_{G,c}^b$ in terms of D_c^b (see 2.8) and the fact that all functors commute with the forgetful functor (theorem 3.4.1(i)).

Similarly for a constructible G-space X, as described in 1.10, we define the full subcategory $D_{G,c}^b(X) \subset D_G^b(X)$ of G-equivariant constructible objects. This category is preserved by all functors for constructible G-maps.

3.7. Integration functors.

Let X be a nice G-space (1.4) and $H \subset G$ be a closed subgroup.

3.7.1. Theorem. *The restriction functor $Res_{H,G} : D_G^b(X) \to D_H^b(X)$ has a right adjoint functor Ind_* and a left adjoint functor $Ind_!$.*

In particular, in case of a trivial subgroup H we have a right and a left adjoint functors to the forgetful functor $For : D_G^b(X) \to D^b(X)$.

Proof. Consider X as an H-space, and denote by Z the induced G-space $Z = ind(X) = G \times_H X$ and by $\pi : Z \to X$ the natural G-map. Denote by $\nu : X \to Z$, $x \mapsto (e,x)$ the natural embedding. By theorem 2.6.3 we have an equivalence of categories $\nu^* : D_B^b(Z) \simeq D_H^b(X)$. As was remarked in 2.6.3 the restriction functor $Res_{H,G} : D_G^b(X) \to D_H^b(X)$ is naturally isomorphic to $\nu^* \cdot \pi^*$.

Let us put $Ind_* = \pi_* \cdot \nu^{*-1} : D_H^b(X) \to D_G^b(X)$. By theorem 3.4.3 this functor is the right adjoint to $Res_{H,G}$.

Consider the equivariant dualizing object $D_\pi = D_{\pi,G} \in D_G^b(Z)$ of the smooth map π (1.4.7, 1.7, 3.4.2). Then we have the canonical isomorphism of functors $\pi^* \simeq D_\pi^{-1} \cdot \pi^!$, where D_π^{-1} stands for the twist functor by D_π^{-1} (1.4.7, 1.5, 3.4.3).

Let us put $Ind_! = \pi_! \cdot D_\pi \cdot \nu^{*-1} : D_H^b(X) \to D_G^b(X)$. Then by theorem 3.4.3 this functor is left adjoint to $Res_{H,G}$.

3.7.2. Proposition. *Let $f : X \to Y$ be a G-map of nice topological. Then the functor Ind_* commutes with functors f_* and $f^!$, and the functor $Ind_!$ commutes with functors $f_!$ and f^*.*

Proof. This follows from 3.4.1, 3.4.3, 1.4.6.

3.7.3. Theorem. *Suppose that the space G/H is ∞-acyclic (1.9), for example contractible. Then the restriction functor $Res_{H,G}$ is fully faithful and its left inverse is the functor Ind_*.*

In particular, if G is an ∞-acyclic group and H is trivial, then the forgetful functor $For : D_G^b(X) \to D^b(X)$ is fully faithful.

Proof. Since the functor Ind_* is the right adjoint to Res it is enough to check that $Ind_* \cdot Res$ is isomorphic to the identity functor. Using the explicit description of the functor Ind_* in the proof of 3.7.1 we see that it amounts to an isomorphism $\pi_* \pi^* \simeq Id$. Since the map $\pi : Z \to X$ is a fibration with the ∞-acyclic fiber G/H it is ∞-acyclic (see 1.9.4). Then the statement follows from 1.9.2 and 3.4.3.

4. Variants.

In sections 2, 3 we have shown that the standard theory of constructible sheaves on topological spaces has a natural extension to the eqivariant situation. The reason for this was the existence of the smooth base change (see 1.8).

There are several theories which are parallel to the theory of constructible sheaves - étale sheaves on algebraic varietes, mixed sheaves on varieties over a finite field, D-modules on complex varieties. In all these situations there exists a smooth base change, so they have natural equivariant extensions. In this section we briefly discuss some of them.

4.1. Let X be a complex algebraic variety, $D_c^b(X)$ be the category of complexes on the topological space X which are constructible with respect to some algebraic stratification. If X is acted upon by a linear algebraic group G we can define the equivariant category $D_{G,c}^b(X)$ in the same way as in section 2. All properties discussed in sections 2 and 3 hold in this case.

4.2. Let X be a complex algebraic variety. Consider the derived category $D^b(D_X)$ of D-modules on X and full subcategories $D_h^b(D_X)$ and $D_{rh}^b(D_X)$ of holonomic and regular holonomic complexes (see [Bo2]).

These categories have functors and properties similar to category $D^b(X)$ in section 1. In particular the smooth base change holds for D-modules (this easily follows from the definition of the functor $f^!$ in [Bo2]).

Let G be a linear algebraic group acting algebraically on X. Using smooth (complex) resolutions as in section 3 we define the equivariant derived categories $D_G^b(D_X), D_{G,h}^b(D_X)$ and $D_{G,rh}^b(D_X)$ and functors between them.

Let $D_{G,c}^b(X)$ be the category described in 4.1. Then the de Rham functor DR, described in [Bo2], establishes the equivariant Riemann-Hilbert correspondence

$$DR : D_{G,rh}^b(D_X) \simeq D_{G,c}^b(X).$$

Remark. A. Beilinson has shown that the category $D^b(D_X)$ can be described directly in terms of D-modules on X in a language analogous to [DV]. We will discuss this interpretation elsewhere.

4.3. Let k be an algebraically closed field. Fix a prime number l prime to $char(k)$. For an algebraic variety X over k we denote by $D_c^b(X)$ the bounded derived category of constructible $\mathbf{Q}_l$-sheaves on X (see [D2]). This category has all functorial properties listed in section 1.

Let G be a linear algebraic group defined over k which acts on X. Then using smooth resolutions as in section 3 we can define the eqivariant derived category $D_{G,c}^b(X)$ and corresponding functors.

5. Equivariant perverse sheaves.

In this section we assume that G is a complex linear algebraic group acting algebraically on a complex variety X. We are interested in the category $D^b_{G,c}(X)$ of equivariant constructible objects on X, defined in 4.1. For a complex variety M we denote by d_M its complex dimension. In the algebraic setting we always assume that the basic ring R is a field of characteristic 0.

5.1 Equivariant perverse sheaves.

We want to define the subcategory of equivariant perverse sheaves $Perv_G(X) \subset D^b_{G,c}(X)$.

Definition. An object $F \in D^b_{G,c}(X)$ lies in the subcategory $Perv_G(X)$ if F_X lies in $Perv(X)$.

It is clear from this definition that all the elementary results about perverse sheaves hold in the equivariant situation. For example, this category is the heart of the "perverse" t-structure on the category $D^b_{G,c}(X)$; in particular it is an abelian category. Every object in $Perv_G(X)$ has finite length and we can describe simple objects in $Perv_G(X)$ in the usual way (see 5.2 below).

Proposition. *(i)* $D(Perv_G(X)) = Perv_G(X)$.

(ii) Let $H \subset G$ be a closed complex normal subgroup, acting freely on X, $B = H \backslash G$. Then the quotient equivalence $q^* : D^b_B(H\backslash X) \to D^b_G(X)$ induces the equivalence $q* : Perv_B(H\backslash X) \to Perv_G(X)[-d_H]$ (2.6.2).

(iii) Let $H \subset G$ be a closed complex subgroup, X - a complex H-variety, $Y = G \times_H X$ and $\nu : X \to Y$ the obvious inclusion. Then the induction equivalence $\nu^* : D^b_G(Y) \to D^b_H(X)$ induces the equivalence $Perv_G(Y) \to Perv_H(X)[d_G - d_H]$ (2.6.3).

Proof. (i) and (ii) are obvious, since all functors commute with the forgetful functor and the category $Perv_G(X)$ is defined in terms of $Perv(X)$.

(iii) Note that it suffices to work only with complex smooth resolutions $P \to X$ (for example using complex Stiefel manifolds as in 3.1). If P is such a resolution, then an object $F \in D^b_{H,c}(X)$ lies in $Perv_H(X)$ iff the object $F(P) \in D^b(\overline{P})$ lies in $Perv(\overline{P})[d_X + d_H - d_P]$. This implies (iii).

5.2. The equivariant intersection cohomology sheaf.

Let $j : V \hookrightarrow X$ be the inclusion of a smooth locally closed irreducible G-invariant subset, and $\mathcal{L} \in Sh_G(V)$ be a G-equivariant local system on V. Consider the intermediate extension $j_{!*}\mathcal{L}[d_V] \in Perv_G(X)$, where d_V is the complex dimension of V ([BBD]). We call this extension the equivariant intersection cohomology sheaf $IC_G(\overline{V}, \mathcal{L})$. In case $\overline{V} = X$ and the local system is trivial $\mathcal{L} = C_V \in Sh_G(V)$ we denote it by $IC_G(X)$. As in the nonequivariant case one can show that simple objects in $Perv_G(X)$ are exactly the intersection cohomology sheaves $IC_G(\overline{V}, \mathcal{L})$, for an irreducible local system $\mathcal{L} \in Sh_G(V)$.

Remark. Note that the quotient and the induction equivalences of proposition 5.1 (ii),(iii) preserve the equivariant intersection cohomology sheaves (up to a shift).

5.3. Decomposition theorem.

An object $F \in D_G^b(X)$ is called semisimple if it is isomorphic to a direct sum of objects $L_i[n_i]$, for some irreducible perverse sheaves $L_i \in D_G^b(X)$.

Theorem. *Let $f : X \to Y$ be a proper G-map of complex algebraic varieties. Let $F \in D_G^b(X)$ be a semisimple object. Then its direct image $H = f_*(F) \in D_G^b(Y)$ is semisimple.*

Proof. Choose a large enough segment $I \subset \mathbf{Z}$ such that $H \in D_G^I(Y)$ (see 2.2). Choose a smooth complex n-acyclic resolution $p : P \to Y$ with $n > |I|$. Then the category $D_G^I(Y)$ is by definition equivalent to the category $D_G^I(X, P)$. By lemma 2.3.2 this category is equivalent to the full subcategory $D^I(\overline{P}|p) \subset D^I(\overline{P})$.

Using the usual decomposition theorem (see [BDD]), we deduce that the object $H \in D^I(\overline{P})$ is semisimple, i.e. is of the form $\oplus H_i$, $H_i \simeq L_i[n_i]$, where L_i are simple perverse sheaves in $D^b(\overline{P})$. Since the subcategory $D^I(\overline{P}|p) \subset D^I(\overline{P})$ is closed with respect to direct summands, all objects H_i lie in this subcategory, which gives the decomposition $H \simeq \oplus H_i$ in $D_G^I(X, P) \simeq D_G^I(X)$. It is clear that every H_i in $D_G^b(X)$ has the form $H_i \simeq L_i[n_i']$, where L_i are irreducible perverse sheaves. This proves the theorem.

6. General inverse and direct image functors Q^*, Q_*.

6.0. Let $\phi : H \to G$ be a homomorphism of topological groups and $f : X \to Y$ be a ϕ-map of of topological spaces (0.1). In this situation we will define functors $Q^* : D_G^b(Y) \to D_H^b(X)$ $(D_G^+(Y) \to D_H^+(X))$ and $Q_* : D_H^+(X) \to D_G^+(Y)$. Many of the functors defined earlier are special cases of these general functors (see 6.6, 6.12 below).

6.1. Assume that the groups H, G satisfy the condition (*) in 2.2.4 (for example, they may be Lie groups). As usual, denote by $Res(X) = Res(X, H)$ and $Res(Y) = Res(Y, G)$ the categories of resolutions of X and Y (2.1.2). We interpret the categories $D_H^b(X)$ and $D_G^b(Y)$ as fibers of the fibered category D^b/T over the functor Φ (2.4.3).

6.2. Definition. Let $P \to X$, $R \to Y$ be resolutions, and $f : P \to R$ be a ϕ-map, such that the diagram

$$
\begin{array}{ccc}
X & \longleftarrow & P \\
\downarrow f & & \downarrow f \\
Y & \longleftarrow & R
\end{array}
$$

is commutative. Then we call resolutions P and R **compatible**.

The following construction produces many compatible resolutions.

Consider the bifunctor

$$\times_f : Res(X) \times Res(Y) \to Res(X), \quad (S, R) \mapsto S \times_X f^0(R).$$

Indeed, $f^0(R) = X \times_Y R$ is naturally an H-space and the projection $f^0(R) \to X$ is an H-map. Note that if $R \to Y$ is n-acyclic then $f^0(R) \to X$ is also n-acyclic. However, $f^0(R)$ is not a free H-space in general, hence not a resolution of X.

We have the obvious map of quotients $\overline{f} : \overline{S \times_f R} \to \overline{R}$, induced by the projection $f : S \times_f R \to R$.

Remarks. 1. If S, R are n-acyclic, then $S \times_f R$ is also such.

2. The trivial resolutions $H \times X \to X$ and $G \times Y \to Y$ are naturally compatible.

3. If $P_1 \to R_1$ and $P_2 \to R_2$ are compatible resolutions then $P_1 \times_X P_2 \to R_1 \times_Y R_2$ are also compatible.

6.3. Definition. A resolution $P \in Res(X)$ is *compatible* (with the map $f : X \to Y$) if it fits into a compatible pair $P \to R$ (6.2). A morphism $P_1 \to P_2$ between compatible resolutions is *compatible* if it fits into a commutative square

$$
\begin{array}{ccc}
P_1 & \longrightarrow & P_2 \\
\downarrow & & \downarrow \\
R_1 & \longrightarrow & R_2
\end{array}
$$

where columns are compatible pairs. Denote by $CRes(X) \subset Res(X)$ the subcategory of compatible resolutions and compatible morphisms.

6.4. Remark. It follows from remarks in 6.2 and proposition 2.4.4 that the category $CRes(X)$ is rich enough to define an object in $D_H^b(X)$.

6.5. Definition of Q_f^*. Let $F \in D_G^b(Y)$ and $P \in CRes(X)$. Let $P \to R$ be a compatible pair of resolutions and $\overline{f} : \overline{P} \to \overline{R}$ be the induced map of quotients. Define

$$Q_f^* F(P) := \overline{f}^* F(R).$$

We must check that the value of $Q_f^* F$ on P is well defined (i.e. is independent of the choice of R), and that $Q_f^* F$ is an object on $D_H^b(X)$.

Let $P \to R'$ be a different compatible pair with the induced map $\overline{f}' : \overline{P} \to \overline{R}'$. Then $P \to R \times_Y R'$ is also a compatible pair, which shows that objects $\overline{f}^* F(R)$ and $\overline{f}^* F(R')$ in $D^b(\overline{P})$ are canonically isomorphic.

Let $\nu : P_1 \to P_2$ be a morphism in $CRes(X)$. We can complete it to a diagram

$$
\begin{array}{ccc}
P_1 & \longrightarrow & P_2 \\
\downarrow & & \downarrow \\
R_1 & \longrightarrow & R_2
\end{array}
$$

as in definition 6.3 above. This shows that objects $\nu^*(Q_f^* F(P_2))$ and $Q_f^* F(P_1)$ are canonically isomorphic. Hence $Q_f^* F$ is a well defined object in $D_H^b(X)$.

6.6. Properties of Q^*.

The properties listed below follow immediately from the definitions.

1. The functor $Q^* : D_G^b(Y) \to D_H^b(X)$ is exact and preserves the t-structure, i.e.

$$Q^* : Sh_G(Y) \to Sh_H(X).$$

2. Let B be another topological group satisfying the condition (*) in 2.2.4. Let $\psi : G \to B$ be a homomorphism and $g : Y \to Z$ be a ψ-map. Then we have a canonical isomorphism of functors

$$Q_f^* \cdot Q_g^* = Q_{gf}^*.$$

3. Suppose that $H = G$ and $\phi = id$. Then Q^* is canonically isomorphic to the inverse image functor f^* in 2.2.4, 3.3.

4. Suppose that $\phi : H \hookrightarrow G$ is an embedding of a subgroup. Then Q^* is canonically isomorphic to the inverse image f^* in 2.6.1. In particular if $X = Y$, we have $Q^* = Res_{H,G}$, and if moreover $H = \{e\}$, then $Q^* = For$.

5. Suppose that $\phi : H \to G$ is surjective with the kernel K, such that X is a free K-space and $q : X \to Y$ is the quotient map by the action of K. Then Q_q^* is naturally isomorphic to the quotient equivalence $q^* : D_G^b(Y) \simeq D_H^b(X)$ in 2.6.2.

6. Suppose that $\phi : H \to G$ is injective, $Y = G \times_H X$ and $\nu : X \to Y$, $x \mapsto (e, x)$. Then Q_ν^* is naturally isomorphic to the induction equivalence $\nu^* : D_G^b(Y) \simeq D_H^b(X)$.

6.7. Let us assume that the groups H, G satisfy the condition (**) in (2.9). Then we can use the definition of D_H^+, D_G^+ in 2.9.1 to define the inverse image functor

$$Q_f^* : D_G^+(Y) \to D_H^+(X),$$

by replacing everywhere in 6.5 the category D^b by D^+.

Alternatively, let $P \to R$ be compatible ∞-acyclic resolutions of X and Y respectively, and $\overline{f} : \overline{P} \to \overline{R}$ be the induced map of quotients. By lemma 2.9.2 the categories $D_H^+(X)$ and $D_G^+(Y)$ are naturally identified as certain full subcategories in $D^+(\overline{P})$ and $D^+(\overline{R})$ respectively. Under this identification we have

$$Q_f^* = \overline{f}^* : D^+(\overline{R}) \to D^+(\overline{P}).$$

All the remarks in 6.6 apply also to D^+.

6.8. Let us define the direct image $Q_{f*} : D_H^+(X) \to D_G^+(Y)$. For simplicity we assume that H, G are Lie groups (and hence satisfy the condition (**) in 2.9).

First we need some local terminology.

Let $p : W \to Z$ be a continuous map of topological spaces. We call p a **good** map if p is locally fibered (1.4.7) with a locally acyclic fiber. This means that for every point $w \in W$ there exist neighbourhoods U of w in W and V of $z = p(w)$ in Z such that $U \simeq F \times V$ where F is acyclic and p is the projection.

Let $GRes(Y) \subset Res(Y, G)$ be the subcategory consisting of good resolutions $r : R \to Y$ (i.e. the map r is good) and good morphisms between them. It follows from our assumptions on the group G that the category $GRes(Y)$ is rich enough to define the category $D_G^+(Y)$ (2.4.4, 2.9.6). Indeed, let N be a locally acyclic free G-space. Then the following diagram of resolutions lies in $GRes(Y)$:

$$G \times Y \leftarrow G \times Y \times N \to Y \times N,$$

(all maps are projections).

Note that if $R_1 \to R_2$ is a morphism in $GRes(Y)$, then the induced map $\overline{R_1} \to \overline{R_2}$ is also good.

6.9. Definition of Q_{f*}. Let $F \in D_H^+(X)$ and $R \in GRes(Y)$. Choose an ∞-acyclic resolution $P \to X$. Consider the compatible pair of resolutions (6.2)

$$P \times_f R \to R$$

and the induced map of quotients

$$\overline{f} : \overline{P \times_f R} \to \overline{R}.$$

We define

$$Q_{f*}F(R) := \overline{f}_* F(P \times_f R).$$

We must check that the value of $Q_{f*}F$ on R is well defined (i.e. does not depend on the choice of P), and that $Q_{f*}F$ is an object in $D_G^+(Y)$.

Let $P' \to X$ be a different ∞-acyclic resolution. Consider the commutative diagram

$$
\begin{array}{ccc}
\overline{(P \times_X P') \times_f R} & \xrightarrow{\ t\ } & \overline{P' \times_f R} \\
\downarrow s & & \downarrow \overline{f'} \\
\overline{P \times_f R} & \xrightarrow{\ \overline{f}\ } & \overline{R}
\end{array}
$$

where all maps are induced by the obvious projections. Put $\overline{F}_P = F(P \times_f R)$, $\overline{F}_{P'} = F(P' \times_f R)$. By definition, we have a canonical isomorphism

$$s^* \overline{F}_P = t^* \overline{F}_{P'}.$$

Note that morphisms s, t are ∞-acyclic, and hence $t_* \cdot t^* = id$, $s_* \cdot s^* = id$ (Proposition 1.9.2(i)). So we have canonical isomorphisms

$$t_* s^* \overline{F}_P = \overline{F}_{P'},$$

$$s_* s^* \overline{F}_P = \overline{F}_P,$$

and therefore a canonical isomorphism

$$\overline{f}_* \overline{F}_P = \overline{f'}_* \overline{F}_{P'},$$

which shows that $Q_{f*}F(R)$ is independent of the choice of P.

Let $g : S \to R$ be a good morphism of resolutions. Consider the pullback diagram

$$
\begin{array}{ccc}
\overline{P \times_f S} & \xrightarrow{\ \overline{g}\ } & \overline{P \times_f R} \\
\downarrow \overline{f} & & \downarrow \overline{f} \\
\overline{S} & \xrightarrow{\ \overline{g}\ } & \overline{R}
\end{array}
$$

In order for $Q_{f*}F$ to be a well defined object in $D_G^+(Y)$ it suffices to show that the base change morphism

$$\overline{g}^* \overline{f}_* \overline{F} \to \overline{f}_* \overline{g}^* \overline{F}$$

is an isomorphism, where $\overline{F} = F(P \times_f R) \in D^+(\overline{P \times_f R})$. Since $\overline{g}$ is a good map this follows from the good base change lemma C.1 proved in Appendix C below. So $Q_{f*}F$ is an object in $D_G^+(Y)$.

6.10. Let us give an alternative description of the functor Q_{f*}. Choose a compatible pair $S \to R$ of ∞-acyclic resolutions (6.2) such that $R \in GRes(Y)$. Let $\overline{f} : \overline{S} \to \overline{R}$ be the induced map of quotients. Recall that categories $D_H^+(X)$ and $D_G^+(Y)$ are canonically identified as certain full subcategories in $D^+(\overline{S})$ and $D^+(\overline{R})$ (2.9.2). After this identification we have

$$Q_{f*} = \overline{f}_* : D^+(\overline{S}) \to D^+(\overline{R})$$

(cf. 6.7 above).

6.11. Example. Let $X = Y = pt$. Let

$$\overline{\phi} : BH \to BG$$

be the map of classifying spaces induced by the homomorphism $\phi : H \to G$. By proposition 2.9.5 the categories $D_H^+(pt)$ and $D_G^+(pt)$ are naturally realized as full subcategories of $D^+(BH)$ and $D^+(BG)$ consisting of complexes with locally constant cohomology. After this identification we have

$$Q^* = \overline{\phi}^* : D^+(BG) \to D^+(BH),$$

$$Q_* = \overline{\phi}_* : D^+(BH) \to D^+(BG).$$

6.12. Properties of Q_*.

1. The functor Q_* is the right adjoint to $Q^* : D_G^+(Y) \to D_H^+(X)$.

Indeed, this is clear from 6.7 and 6.10.

2. Let B be another Lie group and $\psi : G \to B$ be a homomorphism. Let $g : Y \to Z$ be a ψ-map. Then

$$Q_{g*}Q_{f*} = Q_{gf*}.$$

This is clear.

3. Let $K \subset H$ be the kernel of the homomorphism ϕ. Assume that X is a free K-space. For example, ϕ may be injective. The Q_* commutes with the forgetful functor. In particular, if X and Y are nice topological spaces, then Q_* preserves the bounded equivariant category $Q_* : D_H^b(X) \to D_G^b(X)$.

Indeed, let $R \to Y$ be a good ∞-acyclic resolution. Then the H-space $f^0(R) = X \times_Y R$ is free, and hence is an ∞-acyclic resolution of X. So we can use the map $\overline{f} : \overline{f^0(R)} \to \overline{R}$ to define the direct image $Q_* = \overline{f}_*$. Consider the commutative diagram, where both squares are cartesian

$$
\begin{array}{ccccc}
X & \longleftarrow & f^0(R) & \longrightarrow & \overline{f^0(R)} \\
\downarrow f & & \downarrow f & & \downarrow \overline{f} \\
Y & \longleftarrow & R & \longrightarrow & \overline{R}
\end{array}
.
$$

48

Since horizontal arrows are good morphisms (6.8) we conclude by the good base change lemma C1 that Q_* commutes with the forgetful functor.

4. Suppose that $H = G$, $\phi = id$ and $f : X \to Y$ be a G-map of nice topological spaces. Then Q_* preserves the bounded category D_G^b and is naturally isomorphic to f_* in 3.3.

This follows immediately from the property 3 above, since both f_* and Q_* are right adjoint to the same functor $f^* = Q^* : D_G^b(Y) \to D_H^b(X)$.

5. Suppose that $\phi : H \to G$ is injective. Let $X = Y$ and $f = id$. Then Q_* preserves the bounded category D^b and is canonically isomorphic to the integration functor Ind_* (3.7).

This follows immediately from the properties 1,3 above and from 6.6(4).

6. In the situation of the quotient or the induction equivalence (6.6 (5,6)) the functor Q_* preserves the bounded category D^b and is equal to Q^{*-1}.

Indeed, in both cases the functor Q_{f*} preserves the bounded category D^b (see 3 above) and hence is equal to Q_f^{*-1} by 1 above and 6.6(5,6).

7. Some relations between functors.

We establish some relations between the earlier defined functors that we found useful. Roughly speaking, subsections 7.1 - 7.3 contain some commutativity statements, and in 7.4 -7.6 we discuss the behavior of the quotient and the induction equivalences with respect to the Verdier duality.

In this section we assume for simplicity that all groups are Lie groups satisfying the condition (*+) in 3.1 and all spaces are nice (1.4).

7.1. Theorem. (Smooth base change) *Let* $\phi : H \to G$ *be a homomorphism of groups. Consider the pullback diagram*

$$
\begin{array}{ccc}
\tilde{X} & \xrightarrow{\;g\;} & X \\
\downarrow f & & \downarrow f \\
\tilde{Y} & \xrightarrow{\;g\;} & Y
\end{array}
$$

where $g : \tilde{X} \to X$ *(resp.* $g : \tilde{Y} \to Y$ *) is a smooth map of H-spaces (resp. G-spaces) and* f *is a* ϕ*-map. Then*

(i) All functors $\otimes$, *Hom,* f^*, f_*, $f^!$, $f_!$, Q_f^*, Q_{f*} *between categories* D_H^b, D_G^b, *(or* D_H^+, D_H^+*) when defined commute with the smooth base change* g^*.

(ii) The functor g^* *commutes with the Verdier duality* $D : D_H^b(X) \to D_H^b(X)$ *up to the twist by the dualizing object* $D_{g,H} \in D_H^b(\tilde{X})$ *(3.4.2(1)), i.e.*

$$
D \cdot g^* = D_{g,H} \otimes g^* \cdot D.
$$

Proof. This is nothing but the smooth base change (1.8). When ∞-dimensional spaces are involved (functor Q_{f*}) one may use the good base change lemma C1.

7.2. Proposition. *Let* $\phi : H \to G$ *be a homomorphism of groups and* $f : X \to Y$ *be a G-map of G-spaces. Consider* X *as an H-space and* $id : X \to X$ *as a* ϕ*-map. Let* $Q^* = Q_{id}^* : D_G(X) \to D_H(X)$ *be the corresponding inverse image. We have a similar inverse image for the space* Y.

Then all functors $\otimes$, *Hom,* f^*, f_*, $f^!$, $f_!$, D, Q_f^*, Q_{f*} *between categories* D_G, D_H *when defined commute with the inverse image* Q^*. *In particular* Q^* *preserves the dualizing objects.*

For example, if H *is a subgroup in* G *then all the above functors commute with the restriction functor* $Res_{H,G} = Q^*$.

Proof. In this case $Q_f^* = f^*$, $Q_{f*} = f_*$, so it suffices to consider the first seven functors defined in section 3.

Let N be a free smooth G-space, M' be a free smooth H-space. Later on we can assume that N, M' are sufficiently acyclic. Put $M = M' \times N$ - free smooth

H-space. Denote by $P_G = N \times X \to X$, $P_H = M \times X \to X$, $R_G = N \times Y \to Y$, $R_H = M \times Y \to Y$ the corresponding smooth G- and H-resolutions of X and Y. They form the pullback diagram

$$
\begin{array}{ccc}
P_H & \longrightarrow & P_G \\
\downarrow & & \downarrow \\
R_H & \longrightarrow & R_G
\end{array}
$$

where horizontal arrows are projections and vertical ones are induced by f.

This induces the pullback diagram of quotients

$$
\begin{array}{ccc}
\overline{P}_H & \xrightarrow{g} & \overline{P}_G \\
\downarrow \overline{f} & & \downarrow \overline{f} \\
\overline{R}_H & \xrightarrow{g} & \overline{R}_G
\end{array}
$$

where the map g is smooth. By the smooth base change (1.8) the functor g^* commutes with all functors $\otimes$, Hom, $\overline{f}^*$, $\overline{f}_*$, $\overline{f}^!$, $\overline{f}_!$. But g^* represents the inverse image Q^* on the given resolutions. Hence Q^* commutes with the corresponding functors in the equivariant category.

Let us prove that Q^* commutes with the Verdier duality. It suffices to show that Q^* preserves the dualizing objects.

Consider the pullback diagram

$$
\begin{array}{ccc}
\overline{P}_H & \xrightarrow{g} & \overline{P}_G \\
\downarrow p & & \downarrow p \\
M & \xrightarrow{g} & N
\end{array}
$$

Then by definition $D_{X,G}(P_G) = p^! C_{\overline{N}}$, $D_{X,H}(P_H) = p^! C_{\overline{M}}$. Hence again by the smooth base change

$$
g^* D_{X,G}(P_G) = D_{X,H}(P_H)
$$

and so $Q^* D_{X,G} = D_{X,H}$. This proves the proposition.

7.3. Let us prove another base change theorem. Let $\phi : H \to G$ be a homomorphism of groups. Let $G' \subset G$ be a closed subgroup, and $H' = \phi^{-1}(G') \subset H$ be its preimage in H. So we get a commutative diagram of group homomorphisms

$$
\begin{array}{ccc}
H' & \longrightarrow & H \\
\downarrow & & \downarrow \\
G' & \longrightarrow & G
\end{array}
$$

Let $f : X \to Y$ be a ϕ-map. We get a diagram of functors

$$
\begin{array}{ccc}
D_{H'}^+(X) & \xleftarrow{Res_{H',H}} & D_H^+(X) \\
\downarrow Q_{f*} & & \downarrow Q_{f*} \\
D_{G'}^+(Y) & \xleftarrow{Res_{G',G}} & D_G^+(Y)
\end{array}
$$

Theorem. *Assume that ϕ induces an isomorphism $H/H' \simeq G/G'$. Then there is a natural isomorphism of functors*

$$Q_{f*} \cdot Res_{H',H} = Res_{G',G} \cdot Q_{f*}$$

from $D_H^+(X)$ to $D_{G'}^+(Y)$.

Proof. Consider the H-space $H \times_{H'} X$ with the H-map

$$g : H \times_{H'} X \to X, \quad (h, x) \mapsto hx,$$

and the H'-map

$$\nu : X \to H \times_{H'} X, \quad x \mapsto (e, x).$$

Similarly for Y with G and G'.

The restriction functor $Res_{H',H} : D_H^+(X) \to D_{H'}^+(X)$ is isomorphic to the composition of the inverse image

$$g^* : D_H^+(X) \to D_H^+(H \times_{H'} X)$$

with the induction equivalence

$$\nu^* = Q_\nu^* : D_H^+(H \times_{H'} X) \to D_{H'}^+(X).$$

The induction equivalence Q_ν^* commutes with the direct image Q_{f*} because $(Q_\nu^*)^{-1} = Q_{\nu*}$. So it remains to prove that the inverse image g^* commutes with Q_{f*}.

We have a commutative diagram

$$
\begin{array}{ccc}
H \times_{H'} X & \overset{g}{\longrightarrow} & X \\
\downarrow f & & \downarrow f \\
G \times_{G'} Y & \overset{g}{\longrightarrow} & Y
\end{array}
$$

where f is the ϕ-map and g is an H-map (resp. a G-map). Because of our assumption $H/H' = G/G'$ it follows that this is a pullback diagram. Since g is smooth, the assertion follows from 7.1.

7.4. Definition. Let $\phi : H \to G$ be a homomorphism of groups and $f : X \to Y$ be a ϕ-map. Consider Y as an H-space via the homomorphism ϕ and define the dualizing object $D_f = D_{f,H} \in D_H^b(X)$ of the map f as

$$D_f := f^! C_{Y,H}.$$

7.4.1. Lemma. *In the above notations assume that the map f is smooth. Then the inverse image*

$$Q_f^* : D_G^b(Y) \to D_H^b(X)$$

commutes with the Verdier duality up to a twist by the invertible object D_f, i.e.

$$D \cdot Q_f^* = D_f \otimes Q_f^* \cdot D.$$

Proof. Follows immediately from 7.1, 7.2.

7.5. We know that the quotient equivalence q^* commutes with the functors $\otimes$, Hom, f^*, f_*, $f^!$, $f_!$ (3.4.1). Since q^* is the inverse image (and q^{*-1} is the direct image) it also commutes with the functors Q^* and Q_* in the appropriate setting. We claim that q^* commutes with the Verdier duality up to a twist by the invertible dualizing object D_q (7.4).

Namely, let $0 \to H \to G \xrightarrow{\phi} B \to 0$ be an exact sequence of groups. Let X be a G-space which is free as an H-space. Consider $\overline{X} = H \backslash X$ as a B-space and the quotient morphism $q : X \to \overline{X}$ as a ϕ-map. The following proposition is a special case of lemma 7.4.1 above.

7.5.1. Proposition. *The quotient equivalence*

$$q^* : D_B^b(\overline{X}) \simeq D_G^b(X)$$

commutes with the Verdier duality D up to a twist by D_q (7.4). Namely

$$D \cdot q^* = D_q \otimes q^* \cdot D.$$

For a manifold M denote by d_M its dimension.

7.5.2. Proposition. *Assume that in the previous proposition the group G is connected. Then the dualizing object D_q is the constant sheaf shifted by the dimension of H*

$$D_q = C_{X,G}[d_H]$$

and hence

$$D \cdot q^* = (q^* \cdot D)[d_H].$$

That is the quotient equivalence commutes with the Verdier duality up to the shift by d_H.

Proof. Put $\overline{X} = Y$. Consider both X, Y as G-spaces and $f : X \to Y$ as a G-map. Let $P \to Y$ be a smooth resolution of Y and $f^0(P) \to X$ the induced resolution of X. Consider the obvious pullback diagram

$$
\begin{array}{ccc}
f^0(P) & \xrightarrow{q} & \overline{f^0(P)} \\
\downarrow f & & \downarrow \overline{f} \\
P & \xrightarrow{q} & \overline{P}
\end{array}
$$

We must show that

$$\overline{f}^{!} C_{\overline{P}} = C_{\overline{f^0(P)}}[d_H].$$

Since $\overline{f}$ is a smooth map we know that $\overline{f}^{!} C_{\overline{P}} = D_{\overline{f}}$ is invertible (1.4.7, 1.6.1) and locally isomorphic to $C_{\overline{f^0(P)}}[d_H]$. The map q is a smooth fibration with connected fibers ($=G$). Hence it suffices to show that the local system $q^* \overline{f}^{!} C_{\overline{P}}$ is trivial. By the smooth base change

$$q^* \overline{f}^{!} C_{\overline{P}} = f^! q^* C_{\overline{P}} = f^! C_P.$$

But the map f is a principal H-bundle, so $f^! C_P = C_{f^0(P)}[d_H]$ by the following lemma.

7.5.3. Lemma. *Let T be a Lie group and $g : Z \to W$ be a principal T-bundle. Then*

$$g^! C_W = C_Z[d_T].$$

Proof. The fibers of g are orientable ($=T$) and we claim that one can orient the fibers in a compatible way. Indeed, the transition functions in the principal T-bundle are given by the (right) multiplication by elements of T, which preserves the orientation of the fibers.

7.5.4. Corollary. *Under the assumptions of the previous proposition 7.5.2 we have a canonical isomorphism of functors*

$$Q_{q*} \cdot D = (D \cdot Q_{q*})[d_H].$$

Proof. Indeed, $Q_{q*} = q^{*-1}$ (6.12(6)). So apply 7.5.2.

7.6. We know that the induction equivalence (2.6.3) commutes with all functors in section 3 except for the Verdier duality (3.4.1). Here we prove that it commutes with the duality up to a twist by the invertible object D_ν (7.4).

Namely, let $\phi : H \hookrightarrow G$ be an embedding of a closed subgroup and X be an H-space. Consider the induced G-space $Y = G \times_H X$ and the natural ϕ-map $\nu : X \hookrightarrow Y$.

7.6.1. Proposition. *The induction equivalence*

$$Q_\nu^* : D_G^b(Y) \simeq D_H^b(X)$$

commutes with the Verdier duality D up to a twist by the invertible dualizing object $D_\nu \in D^b_H(X)$:

$$D \cdot Q^*_\nu = D_\nu \otimes Q^*_\nu \cdot D.$$

Proof. Since the embedding $\nu : X \hookrightarrow Y$ is relatively smooth (1.4.7(2)), the dualizing object D_ν is indeed invertible.

Since Q^*_ν commutes with Hom it suffices to prove that

$$D_{X,H} = D_\nu \otimes Q^*_\nu D_{Y,G}.$$

By proposition 7.2 $Q^* D_{Y,G} = D_{Y,H}$. So the desired identity is

$$D_{X,H} = D_\nu \otimes \nu^* D_{Y,H},$$

which is the equivariant analogue of 1.4.7(2) (see theorem 3.4.3). This proves the proposition.

7.6.2. Proposition. *Assume that in the previous proposition the group H is connected. Then the dualizing object D_ν is the constant sheaf shifted by the difference of dimensions of H and G*

$$D_\nu = C_{X,H}[d_H - d_G]$$

and hence

$$D \cdot Q^*_\nu = (Q^*_\nu \cdot D)[d_H - d_G].$$

That is the induction equivalence commutes with the Verdier duality up to the shift by $d_H - d_G$.

Proof. Consider the obvious pullback diagram

$$
\begin{array}{ccc}
X & \xrightarrow{\nu} & Y \\
\downarrow p & & \downarrow p \\
pt & \xrightarrow{i} & G/H
\end{array}
$$

where $p : Y \to G/H$ is a locally trivial fibration with fiber X. By the equivariant analogue of 1.4.7(3) we have

$$\nu^! \cdot p^* = p^* \cdot i^!.$$

Hence it suffices to show that

$$i^! C_{G/H,H} = C_{pt,H}[d_H - d_G].$$

Note that G/H is a manifold of dimension $d_G - d_H$. Hence $i^! C_{G/H,H} \in D^b_H(pt)$ is an invertible equivariant sheaf concentrated in degree $d_G - d_H$. But H is connected, hence it is actually constant.

7.6.3. Corollary. *Under the assumptions of the previous proposition 7.6.2 we have a canonical isomorphism of functors*

$$D \cdot Q_{\nu *} = (Q_{\nu *} \cdot D)[d_G - d_H].$$

Proof. Indeed, $Q_{\nu *} = (Q_\nu^*)^{-1}$ (6.12(6)). So apply 7.6.2.

Appendix C.

In this appendix we prove the good base change lemma.

C1. Lemma. *Consider a pullback diagram of continuous maps of topological spaces*

$$
\begin{array}{ccc}
\tilde{X} & \xrightarrow{\ g\ } & X \\
\downarrow f & & \downarrow f \\
\tilde{Y} & \xrightarrow{\ g\ } & Y,
\end{array}
$$

where g is a good map (6.8). Then the base change morphism of functors

$$g^* \cdot f_* \to f_* \cdot g^*$$

is an isomorphism.

Proof. Let $S \in D^+(X)$ and $z \in \tilde{Y}$. We must prove that the induced map on stalks

$$(g^* f_* S)_z \to (f^* g_* S)_z$$

is an isomorphism.

Since g is a good map, there exists a fundamental system of neighbourhoods U of z such that $U = F \times V$, where F is acyclic and $g : U \to V$ is the projection. It suffices to show that over such U we have a quasiisomorphism of complexes

$$(*) \qquad g^* f_* S(U) \simeq f_* g^* S(U).$$

Let us compute both sides in $(*)$. We may assume that the complex S consists of injective sheaves and that $Y = V$, $\tilde{Y} = U$. Since F is connected the cohomology on the left hand side of $(*)$ is

$$H^i(\Gamma(g^* f_* S)) = H^i(\Gamma(f_* S)) = H^i(X, S).$$

On the other hand, since tha map $g : \tilde{X} \to X$ is ∞-acyclic (1.9.4, 1.9.2) the complex $g^* S$ consists of Γ-acyclic sheaves and we have

$$H^i(\Gamma(f_* g^* S)) = H^i(\Gamma(g^* S)) = H^i(X, S).$$

This proves the lemma.

8. Discrete groups and functors.

In this section all groups are assumed to be **discrete**, except in subsection 8.7.

8.0. Let G be a group and X be a G-space. As we mentioned in section 0, a G-equivariant sheaf F on X in this case is simply a sheaf with an action of G which is compatible with its action on X. As usual $Sh_G(X)$ denotes the abelian category of G-equivariant sheaves (of R-modules) on X. This category was studied by A. Grothendieck in [Groth] who showed in particular that $Sh_G(X)$ has enough injectives. He also considered some functorial properties of $Sh_G(X)$ and our discussion here is a variation on the theme of [Groth]. The main point is that we have a natural equivalence of categories

$$D(Sh_G(X)) \simeq D_G(X),$$

where D stands for D^b or D^+ (8.3.1). In other words in the case of a discrete group G the "naive" category $D(Sh_G(X))$ is good enough. Our main objective here is to give a different description of the direct image Q_* in this case (8.4.2) and to study its properties when the action is almost free. In section 9 we apply these results to some actions of algebraic groups.

Notice that given a discrete group G and a G-map $X \to Y$ (of locally compact G-spaces), it is not absolutely clear how to extend the usual functors like $f^!$ to the derived categories $D^b(Sh_G(X))$, $D^b(Sh_G(Y))$. So it may be still useful to work with the equivalent category D_G^b and to use the definition of functors in section 3 above.

8.1. Let $\phi : H \to G$ be a homomorphism of groups and $f : X \to Y$ be a ϕ-map. It induces a natural inverse image functor

$$f^* : Sh_G(Y) \to Sh_H(X).$$

Namely, if $F \in Sh_G(Y)$, then $f^*F \in Sh(X)$ is a sheaf associated to the presheaf

$$f^0F : U \mapsto F(f(U)), \quad U \text{ open in } X.$$

The group G acts naturally on the presheaf f^0F (and hence on the sheaf f^*F) by the formula

$$g : F(f(U)) \xrightarrow{\phi(g)} F(f(V)),$$

where $g(U) = V$.

Assume that the homomorphism ϕ is surjective with the kernel K. Let $S \in Sh_H(X)$. Then the direct image $f_*S \in Sh(Y)$ is naturally an H-equivariant sheaf on Y, considered as an H-space. Hence its subsheaf of K-invariants $(f_*S)^K = f_*^K S$ is naturally a G-equivariant sheaf on Y.

8.1.1. Proposition. *Let $\phi : H \to G$ be a surjective homomorphism of groups with the kernel K. Let X be an H-space which is free as a K-space. Let $Y = K \backslash X$ and*

$f : X \to Y$ be the quotient map. Clearly, Y is a G-space and f is a ϕ-map. Then $f^* : Sh_G(Y) \to Sh_H(X)$ is an equivalence of categories, and the inverse functor is f_*^K.

Proof. One checks immediately that the natural morphisms of functors

$$Id_{Sh_G(Y)} \to f_*^K f^*,$$

$$f^* f_*^K \to Id_{Sh_H(X)}$$

are isomorphisms.

8.2. Let $H \subset G$ be a subgroup and X be a G-space. Consider the natural restriction functor

$$Res_G^H : Sh_G(X) \to Sh_H(X).$$

8.2.1 Definition. Let $S \in Sh_H(X)$. We say that $F \in Sh_G(X)$ is **induced** from S, $F = Ind_H^G(S)$ if
  (a) S is a subsheaf of F,
  (b) $F = \oplus_{s \in G/H} sS$.

Clearly,
$$\mathrm{Hom}_{Sh_G}(Ind_H^G(S), F') = \mathrm{Hom}_{Sh_H}(S, Res_G^H F')$$

for every $F' \in Sh_G(X)$.

8.2.2. Lemma. *For every $S \in Sh_H(X)$ there exists a unique induced sheaf $Ind_H^G(S) \in Sh_G(X)$. The functor*
$$Ind_H^G : Sh_H(X) \to Sh_G(X)$$

is exact.

Proof. Consider the $G \times H$-space $G \times X$ with the action $(g, h)(g', x) = (gg'h^{-1}, hx)$ and the projection $p : G \times X \to X$. Then p^*S is naturally a $G \times H$-equivariant sheaf on $G \times X$. Since $G \times X$ is a free H-space, by proposition 8.1.1 there exists a unique G-equivariant sheaf S' on the G-space $G \times_H X$ such that $p^*S = q^*S'$, where $q : G \times X \to G \times_H X$ is the quotient map by H. Let $m : G \times_H X \to X$ be the G-map $(g, x) \mapsto gx$. Note, that $G \times_H X = \coprod_{s \in G/H}(s, X)$, where (s, X) is homeomorphic to X. Consider the exact functor $m_! : Sh_G(G \times_H X) \to Sh_G(X)$. Then $m_! S' = \oplus_{s \in G/H} sS = Ind_H^G(S)$ is the desired induced sheaf. The uniqueness of the induced sheaf is obvious. This proves the lemma.

8.2.3. Corollary. *The restriction functor Res_G^H (8.2) has a left adjoint exact induction functor Ind_H^G.*

8.2.4. Corollary. *The restriction functor Res_G^H maps injective sheaves to injective. In particular, the forgetful functor $Res_G^{\{e\}} = For : Sh_G(X) \to Sh(X)$ preserves injectives.*

Proof. Indeed, the functor is the right adjoint to the exact functor Ind_H^G.

8.2.5. Corollary. *The functor $Ind_H^G : D(Sh_H(X)) \to D(Sh_G(X))$ is the left adjoint to $Res_G^H : D(Sh_G(X)) \to D(Sh_H(X))$, where D denotes D^b or D^+, and the functors are the trivial extensions of the corresponding exact functors between the abelian categories.*

Proof. This follows immediately from 8.2.3 and 8.2.4.

8.2.6. Proposition. *The induction functor Ind commutes with the inverse image. Namely, let $\phi : G_1 \to G_2$ be a homomorphism and $f : X \to Y$ be a ϕ-map. Let $H_2 \subset G_2$ be a subgroup and $H_1 = \phi^{-1}(H_2) \subset G_1$. Assume that ϕ induces a bijection $G_1/H_1 = G_2/H_2$. Then there is a natural isomorphism o f functors*

$$Ind_{H_1}^{G_1} \cdot f^* \simeq f^* \cdot Ind_{H_2}^{G_2} : Sh_{H_2}(Y) \to Sh_{G_1}(X).$$

Proof. Indeed, let $S \in Sh_{H_2}(Y)$. Then the inverse image $f^*(Ind_{H_2}^{G_2}(S)) \in Sh_{G_1}(X)$ is the induced sheaf $Ind_{H_1}^{G_1}(f^*F)$ (the functor f^* preserves direct sums).

8.3. Let us recall the definition of the functor

$$i : D(Sh_G(X)) \to D_G(X),$$

where D denotes D^b or D^+ (2.5.4).

Let M be a contractible free G-space. Consider the ∞-acyclic resolution p : $P = M \times X \to X$ of X and let $q : P \to \overline{P}$ be the quotient map. Then $D_G(X) \simeq D_G(X, P)$, where the last category is the full subcategory of $D(\overline{P})$ consisting of objects $\overline{F}$ such that $q^*\overline{F}$ comes from X.

Let $S \in D(Sh_G(X))$ be a complex of equivariant sheaves on X. Then $p^*S \in D(Sh_G(P))$, and since P is a free G-space there exists a unique $T \in D(\overline{P})$ such that $q^*T = p^*S$ (lemma 0.3). We put $i(S) = T$.

8.3.1. Theorem. *The above functor $i : D^+(Sh_G(X)) \to D_G^+(X)$ is an equivalence of categories.*

Proof. Using 2.5.3 it suffices to prove that for $F, H \in D^+(Sh_G(X))$

$$\mathrm{Hom}_{D(Sh_G(X))}(F, H) = \mathrm{Hom}_{D_G(X)}(i(F), i(H)).$$

Note that

$$\mathrm{Hom}_{D_G(X)}(i(F), i(H)) = \mathrm{Hom}_{D(\overline{P})}(i(F), i(H)) = \mathrm{Hom}_{D(Sh_G(P))}(p^*F, p^*H).$$

Hence we should prove that

$$(*) \qquad \operatorname{Hom}_{D(Sh_G(X))}(F, H) = \operatorname{Hom}_{D(Sh_G(P))}(p^* F, p^* H).$$

We will reduce $(*)$ to its nonequivariant version, which holds because the map p is ∞-acyclic.

Put $Ind(.) = Ind^G_{\{e\}}(.)$. Assume first that $F = Ind(C_U) \in Sh_G(X)$. Then $p^* F = Ind(C_{p^{-1}(U)})$ (8.2.6). By corollary 8.2.5 the left and right terms in $(*)$ are equal to $\operatorname{Hom}_{D(X)}(C_U, For(H))$ and $\operatorname{Hom}_{D(P)}(C_{p^{-1}(U)}, p^* For(H))$ respectively. But these two groups are equal, since the map p is ∞-acyclic.

Let $F \in Sh_G(X)$. We can find a left resolution of F of the form

$$\ldots \to \oplus_V Ind(C_V) \to \oplus_U Ind(C_U) \to F \to 0.$$

By standard arguments (see [H],7.1) we deduce that $(*)$ holds for $F \in D^b(Sh_G(X))$.

A general $F \in D^+(Sh_G(X))$ can be represented as an inductive limit of bounded complexes

$$F = \lim_{\longrightarrow} \tau_{\leq n} F.$$

Assume that H consists of injective sheaves. Then the complex $Hom^{\cdot}(F, H)$ is the surjective inverse limit

$$Hom^{\cdot}(F, H) = \lim_{\longleftarrow} Hom^{\cdot}(\tau_{\leq n} F, H).$$

Hence the isomorphism $(*)$ for $F \in D^b$ implies the isomorphism for $F \in D^+$. This proves the theorem.

8.4. By the above theorem we may identify the categories $D^+(Sh_G(X))$ and $D^+_G(X)$.

Let $\phi : H \to G$ be a homomorphism of groups anf $f : X \to Y$ be a ϕ-map. Clearly, the inverse image $f^* : D(Sh_G(Y)) \to D(Sh_H(X))$ defined in 8.1 corresponds to the inverse image $Q^* : D_G(Y) \to D_H(X)$ under the identification $D(Sh_G) = D_G$. We want to identify the direct image Q_* (6.9) explicitly when the homomorphism ϕ is surjective.

Assume that $\phi : H \to G$ is surjective, $K = ker(\phi)$. Consider the functor $f^K_* : Sh_H(X) \to Sh_G(Y)$ defined in 8.1. This is a left exact functor (as a composition of two left exact functors), and we denote by Rf^K_* its right derived functor

$$Rf^K_* : D^+(Sh_H(X)) \to D^+(Sh_G(Y)).$$

8.4.1. Proposition. *Let $\phi : H \to G$ be a surjective homomorphism, $K = ker(\phi)$ and $f : X \to Y$ be a ϕ-map. Then the following hold.*

(i) *The functor $f_*^K : Sh_H(X) \to Sh_G(Y)$ is the right adjoint to $f^* : Sh_G(Y) \to Sh_H(X)$.*

(ii) *The functor $Rf_*^K : D^+(Sh_H(X)) \to D^+(Sh_G(Y))$ is the right adjoint to $f^* : D^+(Sh_G(Y)) \to D_H^+(X))$.*

8.4.2. Corollary. *The functor Rf_*^K corresponds to the direct image Q_* under the identification $D^+(Sh_G) = D_G^+$.*

Indeed, both functors are right adjoint to the same functor $f^* = Q^*$.

Proof of 8.4.1. (i) We must show that

$$\operatorname{Hom}_{Sh_H(X)}(f^*F, S) = \operatorname{Hom}_{Sh_G(Y)}(F, f_*^K S),$$

where $F \in Sh_G(Y)$, $S \in Sh_H(X)$.

As in the proof of theorem 8.3.1 we may assume that $F = Ind(C_U)$ for some open subset $U \subset Y$. Then $f^*F = Ind_K^H(C_{f^{-1}(U)})$ (8.2.6). We have

$$\operatorname{Hom}_{Sh_G(Y)}(F, f_*^K S) = \operatorname{Hom}_{Sh(Y)}(C_U, For f_*^K S) = \Gamma(U, f_*^K S) = \Gamma(f^{-1}(U), S)^K,$$

$$\operatorname{Hom}_{Sh_H(X)}(f^*F, S) = \operatorname{Hom}_{Sh_K(X)}(C_{f^{-1}(U)}, Res_H^K S) = \Gamma(f^{-1}(U), S)^K.$$

This proves (i). Now, (ii) follows from (i) and the fact that f_*^K maps injectives to injectives (being the right adjoint to the exact functor f^*).

8.5. In this section we work in the following setup. Let $\phi : H \to G$ be a surjective homomorphism, $K = ker(\phi)$. Let $f : X \to Y$ be a ϕ-map which is the quotient map by the action of K.

8.5.1. Lemma. (i) *Let $F' \in Sh_G(Y)$. The adjunction map*

$$F' \to f_*^K f^* F'$$

is an isomorphism.

(ii) *Let $F \in Sh_H(X)$. The adjunction map*

$$\alpha : f^* f_*^K F \to F$$

is a monomorphism. Moreover α is an isomorphism if and only if F comes from Y, i.e. $F = f^*F'$ for some $F' \in Sh_G(Y)$.

Proof. (see [Groth] in case $G = \{e\}$.)

8.5.2. Corollary. (i) *The functor $f^* : Sh_G(Y) \to Sh_H(X)$ is fully faithful.*

(ii) *For each $F \in Sh_H(X)$ the subsheaf $im(\alpha) \subset F$ is the maximal subsheaf of F that comes from Y.*

8.5.3. Consider the following full subcategories of $Sh_H(X)$:

$$I_K = \{F \in Sh_H(X)| \text{ for each } x \in X, \text{ its stabilizer } K_x \subset K \text{ acts trivially on } F_x\},$$

$$S_K = \{F \in Sh_H(X)| \text{ for each } x \in X, \text{ the stalk } F_x \text{ has no nonzero } K_x\text{-invariants}\}.$$

Remark. The subcategoty I_K is closed under subquotients, but not under extensions in general. The subcategory S_K is closed under extensions, but not under subquotients in general. However, if for each point $x \in X$ the stabilizer K_x is finite and if the basic ring R is a field of characteristic 0, then both subcategories are closed under extensions and subquotients.

8.5.4. Recall the following
Definition. Let B be a group and Z be a B-space. We say that B acts on Z properly discontinuously, if
  (i) the stabilizer B_z of each point $z \in Z$ is finite,
  (ii) each point $z \in Z$ has a neighbourhood V_z such that $bV_z \cap V_z = \emptyset$ if $b \in B$, $b \notin B_z$.

8.5.5. Lemma. *Assume that the group K acts properly discontinuously on X. Let $F \in Sh_H(X)$ and consider the adjunction map*

$$\alpha : f^* f_*^K F \to F.$$

Then for every point $x \in X$ the stalk $im(\alpha)_x \subset F_x$ is equal to K_x-invariants of F_x.
Proof. Fix a point $x \in X$. Since K acts properly discontinuously on X, there exists a fundamental system of neighbourhoods V_x of x with the following properties.
  1. $K_x V_x = V_x$,
  2. $kV_x \cap V_x = \emptyset$ if $k \in K$, $k \notin K_x$.
The stalk $(f^* f_*^K F)_x$ is equal to the limit

$$(f^* f_*^K F)_x = \lim_{V_x} F(f^{-1}(f(V_x)))^K.$$

But $f^{-1}(f(V_x)) = \coprod_{s \in K/K_x} sV_x$ and hence

$$F(f^{-1}(f(V_x)))^K = F(V_x)^{K_x}.$$

Therefore,

$$(f^* f_*^K F)_x = \lim_{V_x} F(V_x)^{K_x} = F_x^{K_x}.$$

This proves the lemma.

8.5.6. Proposition. *Assume that the subgroup $K \subset H$ acts properly discontinuously on X. Then the following hold.*

(i) A sheaf $F \in Sh_H(X)$ belongs to I_K (8.5.3) if and only if it comes from Y.

(ii) Assume that the basic ring of coefficients is a field of characteristic 0. Let $F \in Sh_H(X)$. Consider the canonical exact sequence

$$0 \to F_I \to F \to F_S \to 0,$$

where the map $F_I \to F$ is the adjunction inclusion $\alpha : f^* f_*^K F \to F$ (8.5.1(ii)). Then $F_I \in I_K$, $F_S \in S_K$.

Proof. Let $F \in Sh_H(X)$. Consider the adjunction monomorphism

$$\alpha : f^* f_*^K F \to F.$$

By lemma 8.5.1(ii) we know that it is an isomorphism if and only if F comes from Y. On the other hand, by the previous lemma, it is an isomorphism if and only if $F \in I_K$. This proves (i).

Fix a point $x \in X$. Consider the exact sequence of stalks

$$0 \to F_{I,x} \to F_x \to F_{S,x} \to 0.$$

By the previous lemma, the image of $F_{I,x}$ in F_x coincides with the K_x-invariants. Since K_x is finite and the basic ring R is a field of characteristic 0, the stalk $F_{S,x}$ has no K_x-invariants. This proves (ii).

8.6. Let the basic ring R be a field of characteristic 0.

Let $\phi : H \to G$ be a surjective homomorphism of groups, $K = ker(\phi)$. Let $f : X \to Y$ be a ϕ-map, which is the quotient map by the action of K on X. Assume that K acts on X properly discontinuously.

8.6.1. Theorem. *Under the above assumptions the following hold.*

(i) The functor $f_*^K : Sh_H(X) \to Sh_G(Y)$ is exact.

(ii) $Rf_*^K \cdot f^* \simeq Id_{D^+(Sh_G(Y))}$.

(iii) $Q_* Q^* \simeq Id_{D_G^+(Y)}$.

(iv) Let $F \in Sh_H(X)$. If $F \in S_K$, then $Q_* F = Rf_*^K F = 0$.

(v) The functor f^* induces an equivalence of categories $f^* : Sh_G(Y) \simeq I_K$.

Proof. (ii) and (iii) are equivalent using the identification $D^+(Sh_G) = D_G^+$, $Rf_*^K = Q_*$, $f^* = Q^*$ in 8.4.2. In view of lemma 8.5.1(i) the assertion (i) implies (ii). So it remains to prove (i), (iv), (v).

(i). Let $F \to F'$ be a surjective morphism in $Sh_H(X)$. It suffices to show that $f_*^K F \to f_*^K F'$ is surjective. Fix a point $y \in Y$ and let $x \in X$ be one of its preimages, $f(x) = y$. There exists a fundamental system of neighbourhoods U of y such that $f^{-1}(U) = \coprod_{s \in K/K_x} sV_x$, where V_x is a neighbourhood of x with the following properties:

1. $K_x V_x = V_x$.

2. $kV_x \cap V_x = \emptyset$ if $k \in K$, $k \notin K_x$.

We have

$$f_*^K F(U) = (\prod_{s \in K/K_x} F(sV_x))^K = F(V_x)^{K_x},$$

and similarly

$$f_*^K F'(U) = F'(V_x)^{K_x}.$$

Hence $f_*^K F_y = F_x^{K_x}$, $f_*^K F'_y = F'_x{}^{K_x}$ and it suffices to show that the map $F_x^{K_x} \to F'_x{}^{K_x}$ is surjective. This follows from the surjectivity of $F_x \to F'_x$, since taking K_x-invariants is an exact functor (K_x is a finite group and we work in characteristic 0). This proves (i) and hence also (ii) and (iii).

(iv). It suffices to show that $f_*^K F = 0$ if $F \in S_K$ and then to use (i). So let $F \in S_K$. Using the above argument we find that $f_*^K F_y = F_x^{K_x} = 0$. This proves (iv).

(v). We have $f^* Sh_G(Y) \subset I_K$ (8.5.6(i)), and $f_*^K \cdot f^* = Id_{Sh_G(Y)}$ (8.5.1(i)). Hence it suffices to show that $f_*^K : I_K \to Sh_G(Y)$ is injective on morphisms. But this again follows from the proof of (i) above. This proves (v) and the theorem.

8.7. Let us consider the algebraic situation.

As usual, the basic ring R is assumed to be a field of characteristic 0. Let $\phi : H \to G$ be a surjective homomorphism of groups with a *finite* kernel $K = ker(\phi)$. Let $f : X \to Y$ be a ϕ-map. Assume that X, Y are complex algebraic varieties, f is an algebraic morphism, which is also the quotient map by the action of K.

8.7.1. Theorem. *Under the above assumptions the following hold.*

(i) *The functor* $Q_* : D_H^+(X) \to D_G^+(Y)$ *preserves the t-structure, i.e.* $Q_* : Sh_H(X) \to Sh_G(Y)$.

(ii) $Q_* Q^* \simeq Id_{D_G^+(Y)}$.

(iii) *The functor* Q_* *is exact in the perverse t-structure, i.e.* $Q_* : Perv_H(X) \to Perv_G(Y)$.

(iv) $Q_* IC_H(X) = IC_G(Y)$.

Proof. Since X is a Hausdorff topological space (in the classical topology) and the group K is finite, it acts on X properly discontinuously. Hence (i), (ii) follow from theorem 8.6.1(i),(iii).

(iii). Notice that $f : X \to Y$ is a finite morphism of algebraic varieties and hence $f_* Perv(X) \subset Perv(Y)$.

Let $P \in Perv_H(X)$ be a complex of equivariant sheaves (using the equivalence $D_H^b(X) \simeq D^b(Sh_H(X))$). We have to check that the complex $f_*^K P \in D^b(Sh_G(Y))$ satisfies the support and the cosupport conditions. Namely, let $i : Z \hookrightarrow Y$ be an inclusion of a locally closed subvariety of codimension k. We need to show that

$$(*) \qquad H^s(i^* f_*^K P)_z = 0, \quad s > k$$

$$(**) \qquad H^s(i^! f_*^K P)_z = 0, \quad s < k$$

at the generic point $z \in Z$.

We know (*) and (**) for $f_* P$ instead of $f_*^K P$. The functor i^* commutes with taking K-invariants, which is an exact functor. Hence $H^s(i^* f_*^K P) = (H^s(i^* f_* P))^K$ and (*) holds.

In order to prove (**) we may assume that the complex P consists of injective H-equivariant sheaves. Since functors f_*, f_*^K, For preserve injectives, the complexes $f_* P \in D^b(Sh(Y))$ and $f_*^K P \in D^b(Sh_G(Y))$ also consist of injective sheaves. The functor $i^!$ - taking sections with support in Z - is left exact, hence applicable to the complexes $f_* P$ and $f_*^K P$. It commutes with taking K-invariants and $H^s(i^! f_*^K P) = H^s(i^! f_* P)^K$, so (**) holds. This proves (iii).

(iv) Since $Q_* = f_*^K$ we may assume that $H = K$, $G = \{e\}$. Recall that the intersection cohomology sheaf $IC(W)$ on a stratified pseudomanifold W can be constructed from the constant sheaf C_U on the open stratum U by pushing it forward to the union with smaller and smaller strata and by truncating (see [Bo1]).

Let S be a stratification of Y and $T = f^{-1}(S)$ be the induced stratification of X such that $IC(Y)$ and $IC(X)$ are constructible with respect to S and T. Let $V \subset Y$ be the open stratum on Y and $f^{-1}(V) = U$ be the one on X. Let $C_{U,K} \in Sh_K(U)$ be the constant K-equivariant sheaf on U. Its direct image $f_*^K C_{U,K} = C_V$ is the constant sheaf on V. We claim that constructions of $IC(X)$ and $IC(Y)$ from C_U and C_V commute with the direct image f_*^K. Indeed, the operations of pushing forward (direct image) and truncation obviously commute with the exact functor f_*. The functor $(\ )^K$ of taking K-invariants commutes with the pushforward, and since $(\ )^K$ is exact it also commutes with the truncation. This proves (iv) and the theorem.

9. Almost free algebraic actions.

9.0. In this section we consider almost free actions (only finite stabilizers) of reductive algebraic groups and extend theorem 8.7.1 to this situation. As always in the algebraic setting we assume that the basic ring R is a field of characteristic 0. We denote by d_M the complex dimension of a complex algebraic variety M.

Let $\phi : H \to G$ be a surjective (algebraic) homomorphism of affine reductive complex algebraic groups with the kernel $K = ker(\phi)$. Let X and Y be complex algebraic varieties with algebraic actions of H and G respectively. Let $f : X \to Y$ be an algebraic morphism which is a ϕ-map. Assume that the following hold.

a) The group K acts on X with only finite stabilizers.

b) The morphism f is affine and is the geometric quotient map by the action of K (all K-orbits in X are closed).

9.1. Theorem. *Under the above assumptions the following hold.*

(i) The functor $Q_* : D_H^+(X) \to D_G^+(Y)$ *preserves the t-structure, i.e.* $Q_* : Sh_H(X) \to Sh_G(Y)$. *In particular* Q_* *preserves the bounded category* D^b.

(ii) $Q_*Q^* = Id_{D_G(Y)}$.

(iii) The functor Q_* *preserves the perverse t-structure, i.e.* $Q_* : Perv_H(X) \to Perv_G(Y)[d_K]$.

(iv) $Q_*IC_H(X) = IC_G(Y)[d_K]$.

Proof. Let us first prove the theorem in the absolute case $H = K$, $G = \{e\}$.

All assertions of the theorem are local on Y (at least for the Zariski topology). Hence we may assume that X is affine. By our assumptions all K-orbits are closed. Hence by a fundamental theorem of D. Luna (see [Lu]) at each point $x \in X$ there exists an etale slice. This means the following.

There exists an affine K_x-invariant subvariety $S \subset X$ containing x with the following properties. Consider the following natural diagram

$$
\begin{array}{ccc}
K \times_{K_x} S & \xrightarrow{\tilde{\alpha}} & X \\
\downarrow \tilde{f} & & \downarrow f \\
U & \xrightarrow{\alpha} & Y
\end{array}
$$

where f and $\tilde{f}$ are quotient maps by K, $\tilde{\alpha}$ is the obvious K-map and α is the induced map of the quotients. Then

1. This is a pullback diagram.

2. The morphism α (and hence also $\tilde{\alpha}$) is etale.

Claim. It suffices to prove the theorem for the map $\tilde{f}$ instead of f.

Indeed, by the smooth base change theorem (7.1) the assertions (i),(ii),(iii) for $Q_{\tilde{f}*}$ imply those for Q_{f*}. Suppose we proved (iv) for $Q_{\tilde{f}*}$. Then $\alpha^* \cdot Q_{f*}IC_H(X) = IC(U)$ and hence $Q_{f*}IC_H(X)$ is a simple perverse sheaf on Y. It remains to show that for some smooth open dence subset $V \subset Y$ we have $Q_{f*}IC_H(X)|_V = C_V$. We can choose a smooth V in such a way that $f^{-1}(V) \subset X$ is also smooth and hence $IC_H(X)|_{f^{-1}(V)} = C_{f^{-1}(V)} = Q^* C_V$. But the adjunction map $C_V \to Q_* Q^* C_V$ is an isomorphism by (ii), which is what we need. This proves the claim.

So we may (and will) assume that $U = Y$, $K \times_{K_x} S = X$.

Denote by $\psi : K_x \hookrightarrow K$ the natural embedding.

Consider S as a K_x-space and the embedding $\nu : S \to X$ as a ψ-map. Put $g = f|_S : S \to Y$. We have the following commutative diagram

$$
\begin{array}{ccc}
S & \xrightarrow{\nu} & X \\
\downarrow g & & \downarrow f \\
Y & = & Y
\end{array}
$$

Hence $Q_{g*} : D^+_{K_x}(S) \to D^+(Y)$ is equal to the composition of $Q_{\nu*} : D^+_{K_x}(S) \to D^+_K(X)$ with $Q_{f*} : D^+_K(X) \to D^+(Y)$. But $Q_{\nu*}$ is the inverse functor to the induction equivalence $Q^*_\nu = \nu^* : D^+_K(X) \simeq D^+_{K_x}(S)$ (6.12(6)) and hence it preserves the t-structure and the intersection cohomology sheaves (up to a shift) (5.2). Hence it suffices to prove the theorem for Q_{g*} instead of Q_{f*}. It remains to apply theorem 8.7.1 above. This proves the theorem in the absolute case $H = K$ and $G = \{e\}$.

Let us treat the general case.

Let $P \to Y$ be a smooth complex resolution of the G-space Y. Consider the pullback diagram

$$
\begin{array}{ccc}
P \times_Y X & \longrightarrow & X \\
\downarrow & & \downarrow \\
P & \longrightarrow & Y
\end{array}
$$

where horizontal arrows are smooth H- and G-maps respectively and vertical arrows are ϕ-maps. The statements (i),(ii),(iii) of the theorem are invariant under the smooth base change (7.1). If the fibers of $P \to Y$ are connected (which we can assume) then (iv) is also invariant. Hence it suffices to prove the theorem for the map $P \times_Y X \to P$ instead of $X \to Y$. In other words we may (and will) assume that Y is a free G-space.

Let $q : Y \to \overline{Y}$ be the quotient map by the action of G. Consider the composed morphism $q \cdot f : X \to \overline{Y}$ as a ψ-map, where $\psi : H \to \{e\}$. Then this map qf satisfies the assumptions of the theorem, i.e. it is affine and is the quotient map of X by the action of H, which has only finite stabilizers. So by the absolute case of the theorem which was proved above we know (i)-(iv) for the map qf.

Notice that $Q_{qf*} = Q_{q*} Q_{f*}$ and Q_{q*} is the inverse functor to the quotient equivalence $Q^*_q = q^* : D^+(\overline{Y}) \simeq D^+_G(Y)$ (6.12(6)), which preserves the t-structure and intersection cohomology sheaves (5.2). Hence we deduce that (i)-(iv) hold for Q_{f*} as well. This proves the theorem.

Part II. DG-modules and equivariant cohomology.

The main purpose of the three sections 10,11,12 is to prove theorem 12.7.2 - the detailed algebraic description of the categories $D_G^b(pt)$ and $D_G^+(pt)$ for a connected Lie group G. So we suggest that the reader goes directly to this theorem and, if he understands the statement, he may proceed to next sections which use very little from sections 10-12 besides the mentioned theorem.

In section 10 we review the important language of DG-modules over a DG-algebra. In section 11 we study a very special DG-algebra in terms of which we eventually describe the categories $D_G^b(pt)$ and $D_G^+(pt)$. In section 12 the main theorem 12.7.2 is proved. Unfortunately, the proof is quite technical, mainly for the bounded below category $D_G^+(pt)$.

10. DG - modules.

Our goal is to introduce the homotopy category and the derived category of DG-modules, and to define the derived functors of Hom and $\otimes$.

Most of the general material is contained in $[I\ell]$, but we review the basic definitions for the sake of completeness.

10.1. Definition. A DG-algebra $\mathcal{A} = (A, d)$ is a graded associative algebra $A = \oplus_{i=-\infty}^{\infty} A^i$ with a unit $1_A \in A^0$ and an additive endomorphism d of degree 1 s.t.

$$d^2 = 0$$

$$d(a \cdot b) = da \cdot b + (-1)^{\deg(a)} a \cdot db,$$

and

$$d(1_A) = 0.$$

10.2. Definition. A left DG-module (M, d_M) over a DG-algebra $\mathcal{A} = (A, d)$ (or simply an $\mathcal{A}$–module) is a graded unitary left A-module $M = \oplus_{i=-\infty}^{\infty} M^i$ with an additive endomorphism $d_M : M \to M$ of degree 1 s.t. $d_M^2 = 0$ and

$$d_M(am) = da \cdot m + (-1)^{\deg(a)} a \cdot d_M m$$

for $a \in A, m \in M$. A morphism of DG–modules is a morphism of A–modules of degree zero, which commutes with d.

We will write for short M for (M, d_M) if this causes no confusion.

10.2.1. Denote by $\mathcal{M}_A$ the abelian category of left A-modules.

Note that if $A = A^0$, then an A-module is just a complex of A-modules. In particular $\mathcal{M}_{\mathbb{Z}}$ is the category of complexes of abelian groups.

10.2.2. The **cohomology** $H(M)$ of an A-module M is $H(M) := \operatorname{Ker} d_M / \operatorname{Im} d_M$. Note that $H(M)$ is naturally a graded left module over the graded ring $H(A)$.

10.3. The **translation functor** $[1] : \mathcal{M}_A \to \mathcal{M}_A$ is an automorphism of $\mathcal{M}_A$ s.t.

$$(M[1])^i = M^{i+1}, \quad d_{M[1]} = -d_M$$

and the A-module structure on $M[1]$ is twisted, that is

$$a \circ m = (-1)^{\deg(a)} am,$$

where $a \circ m$ is the multiplication in $M[1]$ and am is the multiplication in M.

10.3.1. Two morphisms $f, g : M \to N$ in $\mathcal{M}_A$ are **homotopic** if there exists a morphism of A-modules (possibly not of A-modules) $M \xrightarrow{s} N[-1]$ s.t.

$$f - g = s d_M + d_N s$$

Null homotopic morphisms $\operatorname{Hot}(M, N)$ form a 2-sided ideal in $\operatorname{Hom}_{\mathcal{M}_A}(M, N)$ and we define the **homotopy category** $\mathcal{K}_A$ to have the same objects as $\mathcal{M}_A$ and morphisms

$$\operatorname{Hom}_{\mathcal{K}_A}(M, N) := \operatorname{Hom}_{\mathcal{M}_A}(M, N) / \operatorname{Hot}(M, N).$$

We now proceed to define the cone of a morphism and the standard triangle in exactly the same way as for complexes of $\mathbb{Z}$-modules.

10.3.2. The **cone** $C(u)$ of a morphism $M \xrightarrow{u} N$ in $\mathcal{M}_A$ is defined in the usual way. Namely, $C(u) = N \oplus M[1]$ with the differential $d_{N \oplus M[1]} = (d_N + u, -d_M)$. We have the obvious diagram

$$M \xrightarrow{u} N \to C(u) \to N[1]$$

in $\mathcal{M}_A$ which is called a **standard triangle**.

10.3.3. An **exact triangle** in $\mathcal{K}_A$ is a diagram isomorphic (in $\mathcal{K}_A$) to a standard triangle above.

10.3.4. Definition. A short exact sequence

$$K \to M \to N$$

of $\mathcal{A}$–modules is called A–split if it splits as a sequence of A-modules.

One can show that an A-split sequence as above can be complemented to an exact triangle

$$K \to M \to N \to K[1]$$

in $\mathcal{K}_\mathcal{A}$.

10.3.5. Proposition. *The homotopy category $\mathcal{K}_\mathcal{A}$ with the translation functor [1] and the exact triangles defined as above forms a triangulated category (see [Ve1]).*

Proof. The proof for complexes of $\mathbb{Z}$-modules applies here without any changes.

10.4. A morphism $M \overset{u}{\to} N$ in $\mathcal{M}_\mathcal{A}$ is a **quasiisomorphism** if it induces an isomorphism on the cohomology $H(M) \overset{\sim}{\to} H(N)$.

10.4.1. The **derived category** $D_\mathcal{A}$ is the localization of $\mathcal{K}_\mathcal{A}$ with respect to quasi-isomorphisms (see [Ve1]).

10.4.2. Lemma. *The collection of quasi-isomorphisms in $\mathcal{K}_A$ forms a localizing system (see [Ve1]).*
Proof. Same as for complexes of $\mathbb{Z}$-modules.

10.4.3. Corollary. *The derived category D_A inherits a natural triangulation from $\mathcal{K}_\mathcal{A}$.*

Proof. Same as for complexes.

Later we will develop the formalizm of derived functors between the derived categories $D_\mathcal{A}$ (see 10.12 below).

10.4.4. Remark. One can check that a short exact sequence

$$0 \to M \to N \to K \to 0$$

in $\mathcal{M}_\mathcal{A}$ defines an exact triangle in $D_\mathcal{A}$.

10.5. As in the case of complexes, the functors $\text{Hom}_{\mathcal{K}_{\mathcal{A}}}(M, \cdot), \text{Hom}_{\mathcal{K}_{\mathcal{A}}}(\cdot, N), \text{Hom}_{D_{\mathcal{A}}}(M, \cdot),$ $\text{Hom}_{D_{\mathcal{A}}}(\cdot, N), H(\cdot)$ from the category $\mathcal{K}_{\mathcal{A}}$ or $D_{\mathcal{A}}$ to the category of graded abelian groups are **cohomological**. That is, they take exact triangles into long exact sequences.

10.6. Right DG–modules. One can develop a similar theory for right DG–modules.

10.6.1. Definition. A right DG–module (M, d_M) over $\mathcal{A} = (A, d)$ is a right graded A-module $M = \oplus_{i=-\infty}^{\infty} M^i$ with an additive endomorphism $d_M : M \to M$ of degree 1, s.t. $d_M^2 = 0$ and

$$d_M(ma) = d_M m \cdot a + (-1)^{\deg(m)} m \cdot da.$$

Denote the category of right DG–modules over $\mathcal{A}$ by $\mathcal{M}_{\mathcal{A}}^r$.

One can either proceed to define the homotopy category $\mathcal{K}_{\mathcal{A}}^r$ and the derived category $D_{\mathcal{A}}^r$ in a way similar to left DG–modules, or simply reduce the study of right modules to that of left modules using the following remark 10.6.3 (the two approaches yield the same result).

10.6.2. For a DG–algebra $\mathcal{A} = (A, d)$ we define its **opposite** $\mathcal{A}^{\text{op}} = (A^{\text{op}}, d)$ to have the same elements and the same differential d, but a new multipliciation $a \circ b$ defined by

$$a \circ b := (-1)^{\deg(a) \cdot \deg(b)} ba,$$

where ba denotes the multiplication in A.

10.6.3. Remark. Let $\mathcal{A}$ be a DG–algebra, $\mathcal{A}^{\text{op}}$ its opposite. Then the categories $\mathcal{M}_{\mathcal{A}}$ and $\mathcal{M}_{\mathcal{A}^{\text{op}}}^r$ are naturally isomorphic. Namely, let $M \in \mathcal{M}_{\mathcal{A}}$ be a *left* $\mathcal{A}$-module. We define on M the structure of a *right* $\mathcal{A}^{\text{op}}$–module as follows

$$m \circ a := (-1)^{\deg(a) \cdot \deg(m)} am.$$

One checks that this establishes an isomorphism of categories $\mathcal{M}_{\mathcal{A}} \overset{\sim}{\leftrightarrow} \mathcal{M}_{\mathcal{A}^{\text{op}}}^r$.

10.7. A DG–algebra is called **supercommutative** if $ab = (-1)^{\deg(a) \cdot \deg(b)} ba$. In other words $\mathcal{A}$ is supercommutative of $\mathcal{A} = \mathcal{A}^{\text{op}}$.

10.8. *Hom.*

Let $M, N \in \mathcal{M}_A$. Define the complex $Hom\,(M, N)$ of $\mathbb{Z}$-modules as follows:
$Hom^n(M, N) := \{$morphisms of A-modules $M \to N[n]\}$; if $f \in Hom^n(M, N)$,
then

$$df = d_N f - (-1)^n f d_M.$$

Note that by definition $Hom_{\mathcal{K}_A}(M, N) = H^0 Hom(M, N)$.

10.8.1. One can check that the bifunctor $Hom(\cdot, \cdot)$ preserves homotopies and defines
an exact bifunctor

$$Hom(\cdot, \cdot) : \mathcal{K}_A^0 \times \mathcal{K}_A \to \mathcal{K}_{\mathbb{Z}}.$$

10.8.2. In case $\mathcal{A}$ is supercommutative the complex $Hom\,(M, N)$ has a natural
structure of a DG-module over $\mathcal{A}$. Namely, for $f \in Hom(M, N)$ put $(af)(m) = af(m)$. In this case $Hom(\cdot, \cdot)$ descends to an exact bifunctor

$$Hom(\cdot, \cdot) : \mathcal{K}_A^0 \times \mathcal{K}_A \to \mathcal{K}_A.$$

10.9. $\otimes_A$.

Let $M \in \mathcal{M}_A^r$, $N \in \mathcal{M}_A$ be a left and right DG-modules. Then the graded
$\mathbb{Z}$-module $M \otimes_A N$ is a complex of abelian groups with the differential

$$d(m \otimes n) = d_M m \otimes n + (-1)^{\deg(m)} m \otimes d_N n$$

We denote this complex by $M \otimes_A N$.

10.9.1. The bifunctor $\otimes_A$ preserves homotopies and descends to an exact bifunctor

$$\otimes_A : \mathcal{K}_A^r \times \mathcal{K}_A \to \mathcal{K}_{\mathbb{Z}}.$$

10.9.2. In case $\mathcal{A}$ is supercommutative the complex $M \otimes_A N$ has a natrual structure
of a DG-module over $\mathcal{A}$. Namely, put $a(m \otimes n) = (-1)^{\deg(a) \cdot \deg(m)} ma \otimes n$. Then
$\otimes_A$ descends to an exact bifunctor

$$\otimes_A : \mathcal{K}_A^r \times \mathcal{K}_A \to \mathcal{K}_A.$$

10.10. If $\mathcal{A}$ is supercommutative we have the following functorial isomorphisms
$(M, N, K \in \mathcal{M}_A)$:

$$M \otimes_A (N \otimes_A K) = (M \otimes_A N) \otimes_A K$$

$$Hom(M, Hom(N, K)) = Hom(M \otimes_A N, K)$$

$$Hom_{\mathcal{M}_A}(M, Hom(N, K)) = Hom_{\mathcal{M}_A}(M \otimes_A N, K)$$

$$Hom_{\mathcal{K}_A}(M, Hom(N, K)) = Hom_{\mathcal{K}_A}(M \otimes_A N, K).$$

10.11. Let $\phi : \mathcal{A} \to \mathcal{B}$ be a **homomorphism** of DG–algebras, that is a unitary homomorphism of graded algebras $\phi : A \to B$ which commutes with the differential. Consider $\mathcal{B}$ as a right DG–module over $\mathcal{A}$ via ϕ.

The assignment $M \mapsto \mathcal{B} \otimes_{\mathcal{A}} M$ defines the **extension of scalars** functor

$$\phi^* : \mathcal{M}_{\mathcal{A}} \to \mathcal{M}_{\mathcal{B}}$$

which descends to the exact functor

$$\phi^* : \mathcal{K}_{\mathcal{A}} \to \mathcal{K}_{\mathcal{B}}.$$

On the other hand, if $N \in \mathcal{M}_{\mathcal{B}}$ we can view N as a left $\mathcal{A}$–module via ϕ. This defines the **restriction of scalars** functor

$$\phi_* : \mathcal{M}_{\mathcal{B}} \to \mathcal{M}_{\mathcal{A}}$$

and the exact functor

$$\phi_* : \mathcal{K}_{\mathcal{B}} \to \mathcal{K}_{\mathcal{A}}.$$

The functors ϕ^* and ϕ_* are **adjoint**. Namely, for $M \in \mathcal{M}_{\mathcal{A}}$, $N \in \mathcal{M}_{\mathcal{B}}$ we have

$$\mathrm{Hom}_{\mathcal{M}_{\mathcal{B}}}(\phi^*(M), N) = \mathrm{Hom}_{\mathcal{M}_{\mathcal{A}}}(M, \phi_*(N))$$

$$\mathrm{Hom}_{\mathcal{K}_{\mathcal{B}}}(\phi^*(M), N) = \mathrm{Hom}_{\mathcal{K}_{\mathcal{A}}}(M, \phi_*(N)).$$

In case $\mathcal{A}$ and $\mathcal{B}$ are supercommutative we have also

$$\mathcal{B} \otimes_{\mathcal{A}} (M \otimes_{\mathcal{A}} N) = (\mathcal{B} \otimes_{\mathcal{A}} M) \otimes_{\mathcal{B}} (\mathcal{B} \otimes_{\mathcal{A}} N)$$

for $M, N \in \mathcal{M}_{\mathcal{A}}$, that is ϕ^* is a tensor functor.

10.12. Derived functors.

Our goal is to define derived functors in the sense of Deligne [D3] of Hom and $\otimes_{\mathcal{A}}$. In order to do that we will construct for each DG–module M a quasiisomorphism $P(M) \to M$, where $P(M)$ is the "bar resolution" of M. Then we show that $P(M)$ can be used for the definition of the derived functors. We use the results of N. Spaltenstein [Sp].

Let us recall the notion of a $\mathcal{K}$–projective complex of $\mathbb{Z}$–modules (see [Sp]).

10.12.1. Definition. Let $C \in \mathcal{M}_{\mathbb{Z}}$. We say that C is $\mathcal{K}$–**projective** if one of the following equivalent properties holds:

(i) For each $B \in M_{\mathbb{Z}}$

$$\mathrm{Hom}_{\mathcal{K}_{\mathbb{Z}}}(C, B) = \mathrm{Hom}_{\mathcal{D}_{\mathbb{Z}}}(C, B)$$

(ii) For each $B \in M_{\mathbb{Z}}$, if $H(B) = 0$, then

$$H(Hom(C, B)) = 0$$

The equivalence of (i) and (ii) is shown in [Sp]. We will repeat the argument in lemma 10.12.2.2 below.

Theorem. *For every complex $B \in M_{\mathbb{Z}}$ there exists a $\mathcal{K}$–projective $C \in M_{\mathbb{Z}}$ and a quasiisomorphism $C \to B$.*

Proof: see [Sp].

10.12.2. We need to extend the definition and the theorem in 10.12.1 to arbitrary DG–modules.

10.12.2.1. Definiton. Let $P \in \mathcal{M}_{\mathcal{A}}$. Then P is called $\mathcal{K}$–**projective** if one of the equivalent conditions in the following lemma holds.

10.12.2.2. Lemma. *Let $P \in \mathcal{M}_{\mathcal{A}}$. Then the following conditions are equivalent:*

(i) $\mathrm{Hom}_{\mathcal{K}_{\mathcal{A}}}(P, \cdot) = \mathrm{Hom}_{\mathcal{D}_{\mathcal{A}}}(P, \cdot)$

(ii) For every acyclic $C \in \mathcal{M}_{\mathcal{D}_{\mathcal{A}}}$ (that is $H(C) = 0$), the complex $Hom(P, C)$ is also acyclic.

Proof: (i) $\Longrightarrow$ (ii). Let C be acyclic. Then $\mathrm{Hom}_{\mathcal{K}_{\mathcal{A}}}(P, C) = \mathrm{Hom}_{\mathcal{D}_{\mathcal{A}}}(P, C) = 0$. But $\mathrm{Hom}_{\mathcal{K}_{\mathcal{A}}}(P, C) = H^0 Hom\ (P, C)$. So $Hom(P, C)$ is acyclic in degree 0. Using the isomorphism $Hom(P, C[i]) = Hom\ (P, C)[i]$ we conclude that $Hom(P, C)$ is acyclic.

(ii) $\Longrightarrow$ (i). By the definition of morphisms in $\mathcal{K}_{\mathcal{A}}$ and $\mathcal{D}_{\mathcal{A}}$ it suffices to prove the following: For a map $s \in \mathrm{Hom}_{\mathcal{K}_{\mathcal{A}}}(T, P)$ that is a quasi-isomorphism, there exists a map

$$t \in \mathrm{Hom}_{\mathcal{K}_{\mathcal{A}}}(P, T), \quad \text{s.t.} \quad s \cdot t = Id_P.$$

Consider the cone of s in $\mathcal{K}_{\mathcal{A}}$:

$$T \xrightarrow{s} P \to C(s).$$

Then by (ii) $\mathrm{Hom}_{\mathcal{K}_{\mathcal{A}}}(P, C(s)) = 0$. Hence from the long exact sequence of $\mathrm{Hom}_{\mathcal{K}_{\mathcal{A}}}(P, \cdot)$ it follows that there exists $t \in \mathrm{Hom}_{\mathcal{K}_{\mathcal{A}}}(P, T)$ s.t. $s \cdot t = Id_P$. This proves the lemma.

10.12.2.3. Remark. One checks directly that the $\mathcal{A}$-module $\mathcal{A}$ is $\mathcal{K}$-projective.

10.12.2.4. Bar construction. For a DG-module $M \in \mathcal{M}_{\mathcal{A}}$ we will now define its bar resolution $B(M) \in \mathcal{M}_{\mathcal{A}}$ together with a quasiisomorphism

$$B(M) \to M.$$

Then we will prove that $B(M)$ is $\mathcal{K}$-projective, hence there are enough $\mathcal{K}$-projective objects in $\mathcal{K}_{\mathcal{A}}$.

So let $M \in \mathcal{M}_{\mathcal{A}}$. Consider M as just a complex of abelian groups $M \in \mathcal{M}_{\mathbb{Z}}$. Let $S_0 = S(M) \xrightarrow{\varepsilon} M$ be its $\mathcal{K}$-projective resolution in $\mathcal{K}_{\mathbb{Z}}$ (which exists by theorem 10.12.1). We may (and will) assume that ε is surjective. Consider the induced $\mathcal{A}$-module $P_0 = \mathcal{A} \otimes_{\mathbb{Z}} S_0$ corresponding to the natural homomorphism $\mathbb{Z} \to \mathcal{A}$.. There is a natural map of $\mathcal{A}$-modules

$$\delta_0 : P_0 \to M, \quad \delta_0(a \otimes s) = a \cdot \varepsilon(s).$$

We claim that δ_0 induces a surjection on the cohomology $H(P_0) \to H(M)$. Indeed, the map $\varepsilon : S_0 \to M$ is a quasiisomorphism and for the cycles $s \in S_0$ and $1 \otimes s \in P_0$ we have $\varepsilon(s) = \delta_0(1 \otimes s)$.

Let $\mathcal{K} = \mathrm{Ker}(\delta_0)$. Then the exact sequence

$$0 \to K \to P_0 \xrightarrow{\delta_0} M \to 0$$

in $\mathcal{M}_{\mathcal{A}}$ induces the exact sequence on cohomology

$$0 \to H(K) \to H(P_0) \to H(M) \to 0$$

We now repeat the preceding construction with K instead of M, etc. This produces a complex of $\mathcal{A}$-modules

$$(*) \qquad \xrightarrow{\delta_{-3}} P_{-2} \xrightarrow{\delta_{-2}} P_{-1} \xrightarrow{\delta_{-1}} P_0 \to 0.$$

Define a new $\mathcal{A}$-module $B(M) = \oplus_{i=-\infty}^{0} P_{-i}[i]$, where the $\mathcal{A}$-module structure on $P_{-i}[i]$ is the *same* as on P_{-i} and the differential

$$d : P_{-i}[i] \to P_{-i}[i] \oplus P_{-i+1}[i-1] \text{ is}$$

$$d(p) = (d_{P_{-i}}(p), (-1)^{\deg(p)} \delta_{-i}(p)).$$

There is an obvious morphism of $\mathcal{A}$-modules $B(M) \xrightarrow{\delta} M$, where $\delta|_{P_0} = \delta_0$ and $\delta|_{P_{-i}} = 0$ for $i > 0$. We call $B(M)$ the **bar resolution** of M, which is justified by the following claim.

10.12.2.5. Claim. $\delta : B(M) \to M$ *is a quasi-isomorphism.*

Indeed, $B(M)$ is the total complex, associated to the double complex (*) of abelian groups. Hence $H(B(M))$ can be computed using the spectral sequence of the double complex (*). The E_1 term is the complex

$$\to H(P_{-2}) \to H(P_{-1}) \to H(P_0) \to 0,$$

which is exact except at P_0 by the construction of $B(M)$. Hence the spectral sequence degenerates at E_2 and $H(B(M)) = E_2 = H(M)$.

10.12.2.6. Proposition. *The $\mathcal{A}$-module $B(M)$ as constructed above is $\mathcal{K}$-projective.*

Proof. We will prove property (ii) of the lemma 10.12.2.2: for an acyclic $\mathcal{A}$-module C the complex $Hom(B(M), C)$ is acyclic. Since $H^0 Hom(B(M), C) = \mathrm{Hom}_{\mathcal{K}_{\mathcal{A}}}(B(M), C)$ it suffices to prove

$$\mathrm{Hom}_{\mathcal{K}_{\mathcal{A}}}(B(M), C) = 0.$$

So let $f : P \to C$ be a morphism of $\mathcal{A}$-modules, where $H(C) = 0$. We will construct a homotopy $h : f \sim 0$, defining h inductively on the increasing sequence of submodules $B_n = \oplus_{i=0}^n P_{-i}[i] \subset B(M)$.

$n = 0$. Recall that $B_0 = P_0 = \mathcal{A} \otimes_{\mathbb{Z}} S_0$ where $S_0 \in \mathcal{M}_{\mathbb{Z}}$ is $\mathcal{K}$-projective. By the adjunction properties in 1.11 the morphism $f|_{P_0} : P_0 \to C$ of $\mathcal{A}$-modules comes from a morphism $g : S_0 \to C$ of $\mathbb{Z}$-modules. But g is homotopic to zero, because S_0 is $\mathcal{K}$-projective. Hence by the same adjunction property there exists a homotopy $h_0 : P_0 \to C[-1]$ s.t. $f|_{P_0} = dh_0 + h_0 d$.

Suppose we have constructed a homotopy $h_{n-1} : B_{n-1} \to C[-1]$ s.t.

$$f|_{B_{n-1}} = dh_{n-1} + h_{n-1}d.$$

We will extend h_{n-1} to a homotopy $h_n : B_n \to C[-1]$. So we need to define h_n on $P_{-n}[n]$.

Let us introduce a local notation. For $M \in \mathcal{M}_{\mathcal{A}}$ (resp. $M \in \mathcal{M}_{\mathbb{Z}}$) we denote by $M[\bar{n}] \in \mathcal{M}_{\mathcal{A}}$ (resp. $M[\bar{n}] \in \mathcal{M}_{\mathbb{Z}}$) the appropriately shifted module, where the differential and the A–module structure (resp. the differential) are the *same* as in M.

Let $K \subset P_{-n+1}$ be the kernel of δ_{-n+1}. Let $\alpha : S(K) \to K$ be the $\mathcal{K}$–projective resolution in $\mathcal{M}_{\mathbb{Z}}$ used in the construction of $B(M)$. Then $P_{-n} = A \otimes_{\mathbb{Z}} S(K)$ and denote by $i : S(K)[\bar{n}] \to P_{-n}[\bar{n}]$, $i(s) = 1 \otimes s$ the map of $\mathbb{Z}$–complexes. Recall that the differential $d_{B(M)}$ acts on $i(s) = 1 \otimes s \in P_{-n}[\bar{n}]$ as

$$d_{B(M)}(i(s)) = 1 \otimes d_{S(K)}(s) + (-1)^{\deg(s)} \alpha(s)$$

Put $\tilde{\alpha}(s) = (-1)^{\deg(s)} \alpha(s)$

Define an additive map $g : S(K)[\bar{n}] \to C$ as follows:

$$g = f \cdot i - h_{n-1} \cdot \tilde{\alpha}.$$

Note that g has degree zero.

Claim. *g is a map of $\mathbb{Z}$–complexes, i.e.*

$$d_C \cdot g = g \cdot d_{S(K)}.$$

Proof.

$$d_C \cdot g = d_C(f \cdot i - h_{n-1} \cdot \tilde{\alpha})$$

$$= d_C \cdot f \cdot i - d_C \cdot h_{n-1} \cdot \tilde{\alpha}$$

$$= f \cdot d_{B(M)} \cdot i - [d_C \cdot h_{n-1} + h_{n-1} \cdot d_{B(M)} - h_{n-1} \cdot d_{B(M)}] \cdot \tilde{\alpha}$$

$$= f[i \cdot d_{S(K)} + \tilde{\alpha}] - f \cdot \tilde{\alpha} + h_{n-1} \cdot d_K \cdot \tilde{\alpha}$$

$$= f \cdot i \cdot d_{S(K)} + h_{n-1} \cdot [-\tilde{\alpha} \cdot d_{S(K)}]$$

$$= [f \cdot i - h_{n-1} \cdot \tilde{\alpha}] d_{S(K)} = g \cdot d_{S(K)}.$$

Since $S(K)$ is $\mathcal{K}$–projective there exists a homotopy of $\mathbb{Z}$–complexes

$$h : S(K)[\bar{n}] \to C[-1], \quad \text{s.t.}$$

$$h d_{S(K)} + d_C h = g$$

We get

$$hd_{S(K)} + d_C h = f \cdot i - h_{n-1} \cdot \tilde{\alpha}$$

$$hd_{S(K)} + h_{n-1}\tilde{\alpha} + d_C h = f \cdot i$$

Now there is a unique map of $\mathcal{A}$–modules $h_n : P_{-n}[\bar{n}] = \mathcal{A} \otimes_{\mathbb{Z}} S(K)[\bar{n}] \to C[-1]$ which extends the homotopy $h : S(K)[\bar{n}] \to C[-1]$. We claim that $h_n d_{B(M)} + d_C h_n = f$ on $P_{-n}[\bar{n}]$, that is h_n is the desired extension of h_{n-1}.

Indeed,

$$(h_n d_{B(M)} + d_C h_n)(a \otimes s) = h_n d_{B(M)}(a \otimes s) + d_C h_n(a \otimes s)$$

$$= h_n(da \otimes s + (-1)^{\deg(a)} a \otimes d_{S(K)} s + (-1)^{\deg(s)+\deg(a)} a\alpha(s))$$

$$+ d_C((-1)^{\deg(a)} ah(s))$$

$$= (-1)^{\deg(a)+1} da \cdot h(s) + (-1)^{\deg(a)+\deg(a)} a\, h(d_{S(K)} s)$$

$$+ (-1)^{\deg(s)+\deg(a)+\deg(a)} a\, h_{n-1}(\alpha(s)) + (-1)^{\deg(a)} da \cdot h(s)$$

$$+ (-1)^{\deg(a)+\deg(a)} ad_C h(s)$$

$$= a[hd_{S(K)} + h_{n-1}\tilde{\alpha} + d_C h](s)$$

$$= af \cdot i(s) = f(a \otimes s).$$

So we have extended the homotopy $h : f \sim 0$ from B_{n-1} to B_n. This completes the induction step and the proof of the proposition.

10.12.2.7 Remark. In case the algebra A contains a field the bar construction can be simplified. Namely, we do not need the intermediate complexes $S(K)$, since every complex over a field is $\mathcal{K}$–projective.

10.12.2.8. Denote by $\mathcal{KP}_A$ the full triangulated subcategory of $\mathcal{K}_A$ consisting of $\mathcal{K}$-projectives. The following corollary follows immediately from 10.12.2.1-6.

10.12.2.9. Corollary. *The localization functor $\mathcal{K}_A \to \mathcal{D}_A$ induces an equivalence of triangulated categories $\mathcal{KP}_A \simeq \mathcal{D}_A$.*

10.12.3. In sections 10.12.2.4–6 above we proved that the category $\mathcal{K}_A$ has enough $\mathcal{K}$-projective objects. This allows us to define the derived functor of $Hom(\cdot, \cdot)$.

10.12.3.1. Definition. For $M, N \in \mathcal{M}_{\mathcal{A}}$ we define the **derived functor**

$$RHom(M, N) := Hom(B(M), N),$$

where $B(M)$ is the bar resolution of M as in 10.12.2.4.

The results in 10.12.2.2–6 above show that $RHom$ is a well defined exact bifunctor

$$RHom : D_{\mathcal{A}}^0 \times D_{\mathcal{A}} \to D_{\mathbb{Z}},$$

which is a right derived functor of Hom in the sense of Deligne ([D3]).

In case $\mathcal{A}$ is supercommutative we get the exact bifunctor

$$RHom : D_{\mathcal{A}}^0 \times D_{\mathcal{A}} \to D_{\mathcal{A}}$$

10.12.4. Next we want to define the derived functor of $\otimes_{\mathcal{A}}$. Let us recall the following definition, which is again due to Spaltenstein [Sp].

10.12.4.1. Definition. A DG–module $P \in \mathcal{M}_{\mathcal{A}}$ is called $\mathcal{K}$–**flat** if the complex $N \otimes_{\mathcal{A}} P$ is acyclic for every acyclic $N \in \mathcal{M}_{\mathcal{A}}^r$.

10.12.4.2. Lemma. *A $\mathcal{K}$-projective complex $S \in \mathcal{M}_{\mathbb{Z}}$ is $\mathcal{K}$-flat.*

Proof. See [Sp].

10.12.4.3. Proposition. *For every $M \in \mathcal{M}_{\mathcal{A}}$ its bar resolution $B(M) \in \mathcal{M}_{\mathcal{A}}$ is $\mathcal{K}$-flat.*

Proof. Let $C \in \mathcal{M}_{\mathcal{A}}^r$ be acyclic. Recall that $B(M)$ is an $\mathcal{A}$–module associated with a complex of $\mathcal{A}$–modules

$$(*) \qquad \ldots \to P_{-2} \xrightarrow{\delta_{-2}} P_{-1} \xrightarrow{\delta_{-1}} P_0 \to 0.$$

Each term P_{-i} is of the form $\mathcal{A} \otimes_{\mathbb{Z}} S$ where S is a $\mathcal{K}$-projective $\mathbb{Z}$–complex. By the previous lemma 10.12.4.2 S is $\mathcal{K}$–flat, so

$$C \otimes_{\mathcal{A}} P_{-i} = C \otimes_{\mathcal{A}} \mathcal{A} \otimes_{\mathbb{Z}} S = C \otimes_{\mathbb{Z}} S$$

is acyclic.

But $C \otimes_{\mathcal{A}} B(M)$ is a complex associated with the double complex

$$\ldots C \otimes_{\mathcal{A}} P_{-2} \xrightarrow{\pm 1 \otimes \delta_{-2}} C \otimes_{\mathcal{A}} P_{-1} \xrightarrow{\pm 1 \otimes \delta_{-1}} C \otimes_{\mathcal{A}} P_0 \to 0,$$

where the "columns" $C \otimes_{\mathcal{A}} P_{-i}$ are acyclic. Hence $C \otimes_{\mathcal{A}} B(M)$ is acyclic. This proves the proposition.

10.12.4.4. Corollary. *A $\mathcal{K}$-projective object in $\mathcal{M}_{\mathcal{A}}$ is $\mathcal{K}$-flat.*

Proof. Let P be $\mathcal{K}$-projective and let $B(P) \overset{\delta}{\to} P$ be its bar resolution. Then δ is a homotopy eqinvalence. Suppose now that $C \in \mathcal{M}_{\mathcal{A}}^r$ is acyclic. Then $C \otimes_{\mathcal{A}} B(P)$ is acyclic by the last proposition. But $C \otimes_{\mathcal{A}} B(P) \overset{1 \otimes \delta}{\to} C \otimes_{\mathcal{A}} P$ is a homotopy equivalence, hence also $C \otimes_{\mathcal{A}} P$ is acyclic. This proves the corollary.

10.12.4.5. Definition. Let $M \in \mathcal{M}_{\mathcal{A}}$, $N \in \mathcal{M}_{\mathcal{A}}^r$. We define the derived functor

$$N \overset{L}{\otimes}_{\mathcal{A}} M := N \otimes_{\mathcal{A}} B(M),$$

where $B(M)$ is the bar resolution as in 10.12.2.4.

The fact that $B(M)$ is $\mathcal{K}$-projective and $\mathcal{K}$-flat implies that $\overset{L}{\otimes}_{\mathcal{A}}$ is a well defined exact bifunctor

$$\overset{L}{\otimes}_{\mathcal{A}} : D_{\mathcal{A}}^r \times D_{\mathcal{A}} \to D_{\mathbb{Z}},$$

which is the left derived functor of $\otimes_{\mathcal{A}}$ in the sense of Deligne ([D3]).

In case $\mathcal{A}$ is supercommutative we get the exact bifunctor

$$\overset{L}{\otimes}_{\mathcal{A}} : D_{\mathcal{A}} \times D_{\mathcal{A}} \to D_{\mathcal{A}}.$$

10.12.5. Let $\phi : \mathcal{A} \to \mathcal{B}$ be a homomorphism of DG–algebras. We now define the derived functor of the extension of scalars functor $\mathcal{B} \otimes_{\mathcal{A}} = \phi^* : \mathcal{K}_{\mathcal{A}} \to \mathcal{K}_{\mathcal{B}}$ to be

$$\mathcal{B} \overset{L}{\otimes}_{\mathcal{A}} = \phi^* : D_{\mathcal{A}} \to D_{\mathcal{B}}$$

We also have the restriction functor

$$\phi_* : D_{\mathcal{B}} \to D_{\mathcal{A}}$$

obtained by restriction of scalars from $\mathcal{B}$ to $\mathcal{A}$. The above functors are adjoint:

$$\mathrm{Hom}_{D_{\mathcal{B}}}(\phi^*(M), N) = \mathrm{Hom}_{D_{\mathcal{A}}}(M, \phi_*(N))$$

for $M \in D_{\mathcal{A}}, N \in D_{\mathcal{B}}$.

10.12.5.1. Theorem. *Let $\phi : \mathcal{A} \to \mathcal{B}$ be a homomorphism of DG–algebras which induces an isomorphism on cohomology $H(\mathcal{A}) \overset{\sim}{\to} H(\mathcal{B})$. Then the extension and the restriction functors*

$$\phi^* : D_{\mathcal{A}} \to D_{\mathcal{B}},$$

$$\phi_* : D_{\mathcal{B}} \to D_{\mathcal{A}}$$

are mutually inverse equivalences of categories.

Proof. Let $M \in D_{\mathcal{A}}$, and $B(M) \overset{\delta}{\to} M$ its bar resolution. Then $\phi_* \cdot \phi^*(M) = \mathcal{B} \otimes_{\mathcal{A}} B(M)$ considered as a left $\mathcal{A}$–module. Define a morphism of functors

$$\alpha : Id_{D_{\mathcal{A}}} \to \phi_* \cdot \phi^*$$

where $\alpha : M \to \mathcal{B} \otimes_{\mathcal{A}} B(M)$ is a composition of δ^{-1} with the map f:

$$f : B(M) \to \mathcal{B} \otimes_{\mathcal{A}} B(M)$$

$$f : e \mapsto 1 \otimes e.$$

To show α is a quasiisomorphism it suffices to show that f is so. But this is immediate since $B(M)$ is $\mathcal{K}$–flat (proposition 10.12.4.3) and hence

$$f = \phi \otimes id : \mathcal{A} \otimes_{\mathcal{A}} B(M) \to \mathcal{B} \otimes_{\mathcal{A}} B(M)$$

is a quasiisomorphism.

Let us define a morphism of functors

$$\beta : \phi^* \cdot \phi_* \to Id_{D_{\mathcal{B}}}.$$

For $N \in D_{\mathcal{B}}$, $\phi^* \cdot \phi_*(N) = \mathcal{B} \otimes_{\mathcal{A}} B(N)$, where $B(N) \overset{t}{\to} N$ is the bar resolution of N considered as an $\mathcal{A}$–module. Put

$$\beta : \mathcal{B} \otimes_{\mathcal{A}} B(N) \to N$$

$$\beta : b \otimes e \mapsto bt(e)$$

We claim that β is a quasiisomorphism. Indeed, consider the commutative diagram

$$B(N) = \mathcal{A} \otimes_{\mathcal{A}} B(N)$$

$$\phi \otimes 1 \downarrow \qquad \searrow \quad 1 \otimes t$$

$$\mathcal{B} \otimes_{\mathcal{A}} B(N) \overset{\beta}{\longrightarrow} \mathcal{A} \otimes_{\mathcal{A}} N = N,$$

where the maps $\phi \otimes 1$ and $1 \otimes t$ are quasiisomorphisms.

This proves the theorem.

10.12.6. If $\mathcal{A}$ is supercommuntative, there are the following functional identities $(M, N, K \in D_{\mathcal{A}})$:

$$M \overset{L}{\otimes}_{\mathcal{A}} (N \overset{L}{\otimes}_{\mathcal{A}} K) = (M \overset{L}{\otimes}_{\mathcal{A}} N) \overset{L}{\otimes}_{\mathcal{A}} K$$

$$RHom\ (M, RHom(N, K)) = RHom(M \overset{L}{\otimes}_{\mathcal{A}}, K)$$

$$\mathrm{Hom}_{D_{\mathcal{A}}}(M, RHom(N, K)) = \mathrm{Hom}_{D_{\mathcal{A}}}(M \overset{L}{\otimes}_{\mathcal{A}} N, K)$$

If in addition $\mathcal{B}$ is supercommutative then $\phi^* : D_{\mathcal{A}} \to D_{\mathcal{B}}$ is a tensor functor, that is

$$\mathcal{B} \overset{L}{\otimes}_{\mathcal{A}} (M \overset{L}{\otimes}_{\mathcal{A}} N) = (\mathcal{B} \overset{L}{\otimes}_{\mathcal{A}} M) \overset{L}{\otimes}_{\mathcal{B}} (\mathcal{B} \overset{L}{\otimes}_{\mathcal{A}} N).$$

11. Categories $D_{\mathcal{A}}^f$, $D_{\mathcal{A}}^+$.

In this work we will be interested in a very special DG–algebra $\mathcal{A} = (A, d = 0)$, where $A = \mathbb{R}[X_1, \ldots, X_n]$ is the commutative polynomial ring, and the generators x_i have various *even* degrees. This is a supercommutative (even commutative) DG–algebra. Denote by $m := (X_1, \ldots, X_n)$ the maximal ideal in A. The algebra A appears as the cohomology ring $H(BG, \mathbb{R})$ of the classifying space BG of a connected Lie group G. Eventually, we will describe the derived category $D_G(pt)$ of G–equivariant sheaves on a point using the derived category $D_{\mathcal{A}}$ (see next section).

11.1. For the remaining part of this section let us fix a DG–algebra $\mathcal{A}$ as above. Consider the following triangulated full subcategories of $\mathcal{M}_{\mathcal{A}}$:

$$\mathcal{M}_{\mathcal{A}}^f = \{M \in \mathcal{M}_{\mathcal{A}} | M \text{ is a finitely generated } A - \text{module}\},$$

$$\mathcal{M}_{\mathcal{A}}^+ = \{M \in \mathcal{M}_{\mathcal{A}} | M^i = 0 \text{ for } i << 0\}.$$

We repeat the construction of the homotopy category and the derived category in section 10 above replacing the original abelian category $\mathcal{M}_{\mathcal{A}}$ by $\mathcal{M}_{\mathcal{A}}^f$ (resp. $\mathcal{M}_{\mathcal{A}}^+$). Denote the resulting categories by $\mathcal{K}_{\mathcal{A}}^f$, $D_{\mathcal{A}}^f$, (resp. $\mathcal{K}_{\mathcal{A}}^+$, $D_{\mathcal{A}}^+$). We have the obvious fully faithful inclusions of categories

$$\mathcal{M}_{\mathcal{A}}^f \subset \mathcal{M}_{\mathcal{A}}^+ \subset \mathcal{M}_{\mathcal{A}},$$
$$\mathcal{K}_{\mathcal{A}}^f \subset \mathcal{K}_{\mathcal{A}}^+ \subset \mathcal{K}_{\mathcal{A}}$$

The following proposition implies that there are similar inclusions of the derived categories (see 11.1.3 below).

11.1.1 Proposition. *Let $M \in \mathcal{M}_{\mathcal{A}}$. Assume that $M \in \mathcal{M}_{\mathcal{A}}^f$ (resp. $M \in \mathcal{M}_{\mathcal{A}}^+$). Then there exists a $\mathcal{K}$-projective $P \in \mathcal{M}_{\mathcal{A}}$ and a quasiisomorphism $P \to M$, such that $P \in \mathcal{M}_{\mathcal{A}}^f$ (resp. $P \in \mathcal{M}_{\mathcal{A}}^+$).*

Proof. Consider the exact sequence of $\mathcal{A}$-modules

$$(*) \qquad\qquad 0 \to \text{Ker} d_M \to M \to M/\text{Ker} d_M \to 0,$$

where the $\mathcal{A}$-modules $\text{Ker} d_M$ and $M/\text{Ker} d_M$ have zero differential and belong to $\mathcal{M}_{\mathcal{A}}^f$ (resp. $\mathcal{M}_{\mathcal{A}}^+$). The sequence $(*)$ is an exact triangle in $D_{\mathcal{A}}$. Since $\mathcal{K}$-projective objects form a triangulated subcategory in $\mathcal{K}_{\mathcal{A}}$, we may assume that the differential d_M in the module M is zero.

Let

$$0 \to P_{-n} \xrightarrow{d} \ldots \xrightarrow{d} P_0 \xrightarrow{\varepsilon} M \to 0$$

be a graded resolution of M (as an A-module) by finitely generated (resp. bounded below) projective (hence free) A-modules. This resolution defines an A-module

$P = \oplus P_{-i}[i]$ with the differential $d : P_{-i}[i] \to P_{-i+1}[i-1]$ and a quasiisomorphism $\varepsilon : P \xrightarrow{\sim} M$ ($\varepsilon|_{P_{-i}[i]} = 0$, $i > 0$). Clearly $P \in \mathcal{M}_{\mathcal{A}}^f$ (resp. $P \in \mathcal{M}_{\mathcal{A}}^+$).

Claim. *The $\mathcal{A}$-module P is $\mathcal{K}$-projective.*

Proof of the claim. One immitates the proof of the corresponding statement for complexes (see [Hart]). Namely, the following statement is easily verified: Let $C \in \mathcal{M}_{\mathcal{A}}$, s.t. $H(C) = 0$, then $\mathrm{Hom}_{\mathcal{K}_{\mathcal{A}}}(P, C) = 0$.

This proves the proposition.

11.1.2. Consider the full subcategory $\mathcal{KP}_{\mathcal{A}} \subset \mathcal{K}_{\mathcal{A}}$ consisting of $\mathcal{K}$-projectives and denote by $\mathcal{KP}_{\mathcal{A}}^f \subset \mathcal{KP}_{\mathcal{A}}$ the full subcategory consisting of objects $M \in \mathcal{M}_{\mathcal{A}}^f$. We have a natural commutative diagram of functors

$$
\begin{array}{ccc}
\mathcal{KP}_{\mathcal{A}}^f & \xrightarrow{a} & D_{\mathcal{A}}^f \\
\downarrow c & & \downarrow d \\
\mathcal{KP}_{\mathcal{A}} & \xrightarrow{b} & D_{\mathcal{A}}.
\end{array}
$$

The functor b is an equivalence (10.12.2.9). It follows from the above proposition that the functor a is also an equivalence. Hence d is fully faithful. Similarly for $\mathcal{KP}_{\mathcal{A}}^+$ and $D_{\mathcal{A}}^+$. So we proved the following

11.1.3. Corollary. *The natural functor $D_{\mathcal{A}}^f \to D_{\mathcal{A}}$ (resp. $D_{\mathcal{A}}^+ \to D_{\mathcal{A}}$) is fully faithful.*

11.1.4. Actually the proof of 11.1.1 gives more. Namely, the $\mathcal{K}$-projective DG-module P constructed in this proof is an iterated extension of finite (shifted) direct sums of $\mathcal{A}$ (resp. of bounded below direct sums $\oplus_{i > \mu}(\oplus \mathcal{A}[-i])$) in case of the category $\mathcal{M}_{\mathcal{A}}^f$ (resp. $\mathcal{M}_{\mathcal{A}}^+$). This shows that the triangulated category $\mathcal{KP}_{\mathcal{A}}^f$ (resp. $\mathcal{KP}_{\mathcal{A}}^+$) is generated by $\mathcal{A}$ (resp. by bounded below direct sums $\oplus_{i > \mu}(\oplus \mathcal{A}[-i])$). Let us write for short $\oplus^+ \mathcal{A}[-i]$ for $\oplus_{i > \mu}(\oplus \mathcal{A}[-i])$. Using the equivalence a in 11.1.2 we obtain

11.1.5. Corollary. *The triangulated category $D_{\mathcal{A}}^f$ (resp. $D_{\mathcal{A}}^+$) is generated by $\mathcal{A}$ (resp. by bounded below direct sums $\oplus^+ \mathcal{A}[-i]$).*

11.1.6. The categories $\mathcal{M}_{\mathcal{A}}^f$, $\mathcal{M}_{\mathcal{A}}^+$ (resp. $\mathcal{K}_{\mathcal{A}}^f$, $\mathcal{K}_{\mathcal{A}}^+$) are closed under the tensor product $\otimes_{\mathcal{A}}$, and $\mathcal{M}_{\mathcal{A}}^f$, $\mathcal{K}_{\mathcal{A}}^f$ are closed under $Hom(\cdot, \cdot)$. Using the proposition 11.1.1 we can define the derived functors of $Hom(\cdot, \cdot)$ and $\otimes_{\mathcal{A}}$ on the categories $D_{\mathcal{A}}^f, D_{\mathcal{A}}^+$, using the $\mathcal{K}$-projective resolutions (see definitions 10.12.3,4). So we obtain the exact functors

$$
RHom(\cdot, \cdot) : D_{\mathcal{A}}^f \times D_{\mathcal{A}}^f \to D_{\mathcal{A}}^f
$$

$$RHom(\cdot,\cdot) : D_{\mathcal{A}}^f \times D_{\mathcal{A}}^+ \to D_{\mathcal{A}}^+$$

$$\cdot \overset{L}{\otimes} \cdot : D_{\mathcal{A}}^f \times D_{\mathcal{A}}^f \to D_{\mathcal{A}}^f$$

$$\cdot \overset{L}{\otimes} \cdot : D_{\mathcal{A}}^+ \times D_{\mathcal{A}}^+ \to D_{\mathcal{A}}^+$$

Clearly, all relations in 10.12.6 hold for the above functors.

Our main interest lies in the category $D_{\mathcal{A}}^f$. We proceed to define some additional structures on $D_{\mathcal{A}}^f$.

11.2. Duality.

11.2.1. Definition. The $\mathcal{A}$-module $D_{\mathcal{A}} := \mathcal{A} \in D_{\mathcal{A}}^f$ is called the **dualizing** module.

Next we define the **duality** functor

$$D : D_{\mathcal{A}}^f \to D_{\mathcal{A}}^f$$

$$D(M) := RHom(M, D_{\mathcal{A}})$$

11.2.2. Proposition. $D^2 = Id$.

11.2.2.1. Lemma. *For a $\mathcal{K}$-projective $P \in \mathcal{K}_{\mathcal{A}}^f$ the $\mathcal{A}$-module $Hom(P, D_{\mathcal{A}})$ is also $\mathcal{K}$-projective.*

Proof. Since the category of $\mathcal{K}$-projectives is generated by $\mathcal{A}$ (see 11.1.4), it suffices to prove the lemma if $P = \mathcal{A}$ in which case it is obvious.

Proof of proposition. Let $M \in D_{\mathcal{A}}^f$ be $\mathcal{K}$-projective. Then using the above lemma we have

$$DD(M) = Hom(Hom(M, D_A), D_A).$$

Define a map of $\mathcal{A}$-modules

$$\alpha : M \to DD(M) , \ \alpha(m)(f) = (-1)^{\deg(m)\deg(f)} f(m)$$

We claim that α defines an isomorphism of functors $Id \overset{\sim}{\to} D^2$. By corollary 11.1.5 above it suffices to prove that α is an isomorphism in case $M = \mathcal{A}$, which is obvious. This proves the proposition.

11.3. Relations with Ext_A and Tor_A.

Let us point out some relations between the operations $RHom$ and $\overset{L}{\otimes}_A$ in $D_{\mathcal{A}}^f$ and the operations Ext_A and Tor_A in the category Mod_A of graded $\mathcal{A}$-modules.

Let $M, N \in \mathcal{M}_{\mathcal{A}}^f$ be $\mathcal{A}$-modules with **zero differentials**. Recall the construction of a $\mathcal{K}$-projective resolution P of M as in the proof of proposition 11.1.1. We considered a projective resolution

$$(*) \qquad\qquad 0 \to P_{-n} \xrightarrow{d} \ldots \to P_{-1} \xrightarrow{d} P_0 \xrightarrow{\varepsilon} M \to 0$$

of M in the category Mod_A. Then the $\mathcal{A}$ - module P was defined as $P = \oplus_i P_{-i}[i]$ with the differential $d : P_{-i}[i] \to P_{-i+1}[i-1]$ and the quasiisomorphism $\varepsilon :$ $P \xrightarrow{\sim} M$ $(\varepsilon(P_{-i}[i]) = 0$, for $i > 0)$.

By definition,

$$(1) \qquad\qquad RHom(M, N) = Hom(P, N).$$

On the other hand, the complex of A-modules

$$(2) \qquad\qquad \ldots \to \mathrm{Hom}_A(P_{-i}, N) \to \mathrm{Hom}_A(P_{-i-1}, N) \to \ldots$$

computes the modules $\mathrm{Ext}_A^i(M, N)$

Comparing (1) and (2) we find

11.3.1. Proposition. *If $M, N \in D_{\mathcal{A}}^f$ have zero differentials, then*

$$(i) \qquad\qquad H(RHom(M, N)) = \oplus_i Ext_A^i(M, N)[-i].$$

In particular, if $Ext_A^i(M, N) = 0$, $i \neq k$, then the $\mathcal{A}$-module $RHom(M, N)$ is quasi-isomorphic to its cohomology $H(RHom(M, N))$ and hence

$$RHom(M, N) = Ext_A^k(M, N)[-k].$$

In case $N = D_A = A$ the last equality becomes

$$D(M) = Ext_A^k(M, A)[-k],$$

which shows the close relation between the duality in $D_{\mathcal{A}}^f$ and the coherent duality in Mod_A.

(ii) There exists a natural morphism in $D_{\mathcal{A}}^f$

$$RHom(M, N) \to Ext_A^n(M, N)[-n],$$

which induces a surjection on the cohomology (the differential in the second A-module is zero).

In the previous notations we also have

$$(3) \qquad\qquad M \overset{L}{\otimes}_A N = P \otimes_A N.$$

On the other hand the complex of A-modules

(4) $$\ldots \to P_{-i} \otimes_A N \to P_{-i+1} \otimes_A N \to \ldots$$

computes the modules $Tor_A^i(M, N)$. Comparing (3) and (4) we find

11.3.2. Proposition. *If* $M, N \in D_A^f$ *have zero differentials, then*

(i) $$H(M \overset{L}{\otimes}_A N) = \oplus_i Tor_A^i(M, N)[i].$$

In particular, if $Tor_A^i(M, N) = 0$, $i \neq k$, *then the* A-*module* $M \overset{L}{\otimes}_A N$ *is quasiisomorphic to its cohomology* $H(M \overset{L}{\otimes}_A N)$ *and hence*

$$M \overset{L}{\otimes}_A N = Tor_A^k(M, N)[k].$$

(ii) *There exists a natural morphism in* D_A^f

$$M \overset{L}{\otimes}_A N \to M \otimes_A N,$$

which induces a surjection on the cohomology (the differential in the second module is zero).

11.3.3. For a given $M \in D_A$ it is useful to know if M is quasiisomorphic to its cohomology, i.e. if $M \simeq (H(M), d = 0)$ (see, for example, the previous propositions 11.3.1, 11.3.2).

Proposition. *Let* $M \in D_A$ *be such that the* A-*module* $H(M)$ *has cohomological dimension 0 or 1. Then* $M \simeq (H(M), d = 0)$.

Proof. Choose $\{c_i\} \subset \text{Ker} d_M \subset M$ such that $\{c_i\}$ generate the cohomology $H(M)$ as on A-module. Let $P_0 = \oplus_i A c_i$ be the free A-module on generators c_i with the natural map of A-modules

$$\varepsilon : P_0 \to \text{Ker} d_M$$

$$c_i \mapsto c_i.$$

Let $P_{-1} \subset P_0$ be the kernel of the composed surjective map

$$P_0 \overset{\varepsilon}{\to} \text{Ker} d_M \to \text{Ker} d_M / \text{Im} d_M = H(M).$$

By our assumption P_{-1} is a free A-module, and hence

$$0 \to P_{-1} \overset{d}{\to} P_0 \to H(M) \to 0$$

is a projective resolution of the A-module $H(M)$.

As in the proof of 11.1.1 consider the $\mathcal{K}$-projective $\mathcal{A}$-module $P = P_0 \oplus P_{-1}[1]$ with the differential $d : P_{-1}[1] \to P_0$. Then P is quasiisomorphic to $H(M)$ (from the exact sequence above). Let us construct a quasiisomorphism $P \overset{\sim}{\to} M$, which will prove the proposition.

We already have the map $P_0 \overset{\varepsilon}{\to} M$. It remains to find a map of A-modules $\varepsilon' : P_{-1} \to M$ which makes the following diagram commutative

$$
\begin{array}{ccc}
P_{-1} & \overset{\varepsilon'}{\to} & M \\
\downarrow d & & \downarrow d_M \\
P_0 & \overset{\varepsilon}{\to} & M.
\end{array}
$$

Such a map exists since $\varepsilon \circ d(P_{-1}) \subset Im d_M$ and P_{-1} is a projective A-module. This proves the proposition.

11.4. t-structure.

Let us recall the definition of a t-category in [BBD].

11.4.1. Definition. A t-**category** is a triangulated category D together with two full subcategories $D^{\leq 0}, D^{\geq 0}$ s.t. if $D^{\leq n} := D^{\leq 0}[-n]$ and $D^{\geq n} := D^{\geq 0}[-n]$, then

(i) For $X \in D^{\leq 0}, Y \in D^{\geq 1}, \mathrm{Hom}_D(X,Y) = 0$

(ii) $D^{\leq 0} \subset D^{\leq 1}$ and $D^{\geq 0} \supset D^{\geq 1}$

(iii) For $X \in D$ there exists an exact triangle $A \to X \to B$ s.t. $A \in D^{\leq 0}, B \in D^{\geq 1}$.

Let us introduce a t-structure an $D_{\mathcal{A}}^f$.

11.4.2. Definition.

$D_{\mathcal{A}}^{f, \geq 0} := \{M \in D_{\mathcal{A}}^f |$ there exists $N \in D_{\mathcal{A}}^f$ quasiisomorphic to M such that $N^i = 0, i < 0\}$.

$D_{\mathcal{A}}^{f, \leq 0} := \{N \in D_{\mathcal{A}}^f | \mathrm{Hom}_{D_{\mathcal{A}}^f}(N, M) = 0$ for all $M \in D_{\mathcal{A}}^{f, \geq 1}\}$.

11.4.3. Theorem. *The triple* $(D_{\mathcal{A}}^f, D_{\mathcal{A}}^{f, \geq 0}, D_{\mathcal{A}}^{f, \leq 0})$ *is a* t-*category.*

Proof. The properties (i),(ii) of definition 11.4.1 are obvious. In order to prove (iii) we need some preliminaries. The proof will be finished in 11.4.11 below.

11.4.4. Definition. Let $N \in D_{\mathcal{A}}^f$ be a free A-module. Denote by $rk_A N$ its rank as an A-module. Let now $M \in D_{\mathcal{A}}^f$ be arbitrary. Then we define the **rank** of M as follows

$$
rk M := \min_P \{rk_A P | P \text{ is } \mathcal{K} - \text{projective, free as}
$$

$$
\text{an } A\text{-module, quasiisomorphic to } M\}.
$$

11.4.5. Remark. The function $rk(M)$ satisfies the "triangle inequality". Namely, if $M \to N \to K$ is an exact triangle in $D_{\mathcal{A}}^f$, then $rkN \leq rkM + rkK$.

11.4.6. Definition. Let $P \in D_{\mathcal{A}}^f$ be a $\mathcal{K}$-projective, free as an A-module. Then P is called a **minimal $\mathcal{K}$-projective**, if $rk_A P = rkP$.

11.4.6.1. Let $P = (\oplus A[i], d_P)$ be a free A-module. Define the following A-submodules of P:

$$P_{\leq i} = \oplus_{j \geq -i} A[j]$$

$$P_{\geq i} = \oplus_{j \leq -i} A[j],$$

so that $P = P_{\leq i} \oplus P_{\geq i+1}$.

11.4.7. Lemma. *Let (P, d_P) be a $\mathcal{K}$-projective $\mathcal{A}$-module, free as an A-module. Then the following statements are equivalent.*

(i) $d_P(P) \subset mP$;

(ii) P is a minimal $\mathcal{K}$-projective $\mathcal{A}$-module;

(iii) for all k, $P_{\leq k}$ is an $\mathcal{A}$-submodule of P, i.e. $d_P(P_{\leq k}) \subset P_{\leq k}$.

Proof. Let $e_1, \ldots, e_n$ be a graded A-basis of P, s.t. $\deg(e_{i+1}) \geq \deg(e_i)$. Then the differential d_P is an endomorphism of P given by a matrix $M = (a_{ij})$, where

$$d_P(e_j) = \sum_{i=1}^{n} a_{ij} e_i.$$

Then clearly, (i)$\Leftrightarrow$ $a_{ij} \in m$, $\forall i, j$ $\Leftrightarrow$ M is upper triangular $\Leftrightarrow$(iii).

(i)$\Rightarrow$(ii). Assume that P satisfies (i). Consider the complex of $\mathbb{R}$-vector spaces

$$\mathbb{R} \overset{L}{\otimes}_A P = P/mP.$$

By our assumption this complex has zero differential. Hence P is minimal.

(ii)$\Rightarrow$(i). Induction on the A-rank of P.

Suppose that $d_P(e_1) = 0$. Then we have a short exact sequence of $\mathcal{A}$- modules

$$Ae_1 \to P \to P/Ae_1,$$

where all modules are free. This sequence is A-split, hence defines an exact triangle in $\mathcal{K}$ (10.3.4). The first two terms are $\mathcal{K}$-projective, hence the third one is also such. Moreover, from the triangle inequality (11.4.5) it follows that P/Ae_1 is minimal. By the induction hypothesis (i) holds for P/Ae_1. Hence it also holds for P.

Now suppose that $d_P(e_1) \neq 0$. Then $d_P(e_1)$ is an $\mathbb{R}$-linear combination of e_i's. We may (and will) assume that $d_P(e_1) = e_i$ for some i. Denote by E the $\mathcal{A}$-submodule of P spanned by e_1, e_i. Note that E is $\mathcal{K}$-projective as the cone of the identity morphism $id : Ae_i \to Ae_i$ and $H(E) = 0$. Consider the short exact sequence of $\mathcal{A}$-modules

$$E \to P \to P/E,$$

where all modules are free. This sequence is A-split, hence defines an exact triangle in $\mathcal{K}_{\mathcal{A}}$ (10.3.4). The first two terms are $\mathcal{K}$-projective, hence the third one is also such. The map $P \to P/E$ is a quasiisomorphism, which contradicts the minimality of P. This proves the lemma.

11.4.8. Lemma. *Let P be a minimal $\mathcal{K}$-projective. Then the $\mathcal{A}$-submodules $P_{\leq k} \subset P$ (11.4.7) are $\mathcal{K}$-projective for all k. Hence also $P/P_{\leq k}$ are $\mathcal{K}$-projective and actually $P_{\leq k}$, $P/P_{\leq k}$ are minimal.*

Proof. Induction on k.

11.4.9. Remark. Let $M \in D_{\mathcal{A}}^f$. Let $e_1, \ldots, e_n$ be a graded A-basis for a minimal $\mathcal{K}$-projective module P quasiisomorphic to M. The previous lemma implies that there is an isomorphism of graded $\mathbb{R}$-vector spaces

$$H(\mathbb{R} \overset{L}{\otimes}_{\mathcal{A}} M) = \oplus \mathbb{R} e_i.$$

In particular

$$rkM = dim_{\mathbb{R}} H(\mathbb{R} \overset{L}{\otimes}_{\mathcal{A}} M).$$

11.4.10. Proposition. *Let $P, Q \in \mathcal{M}_{\mathcal{A}}$ be two minimal $\mathcal{K}$-projectives. If P, Q are quasiisomorphic then they isomorphic.*

Proof. Let $a : P \to Q$ be a quasiisomorphism. Since P is $\mathcal{K}$-projective, a is an actual morphism of modules. Applying the functor $\mathbb{R} \otimes_{\mathcal{A}} \cdot$ we find that a induces an isomorphism of the vector spaces

$$P/mP = Q/mQ.$$

Hence a is an isomorphism by the Nakayama lemma.

11.4.11. Now we can finish the proof of theorem 11.4.3. Let $M \in D_{\mathcal{A}}^f$. Let P be the minimal $\mathcal{K}$-projective quasiisomorphic to M (which is unique by 11.4.10). By lemma 11.4.7 the $\mathcal{A}$-submodule $P_{\leq 0} \subset P$ is actually an $\mathcal{A}$-submodule. Consider the exact triangle

$$P_{\leq 0} \to P \to P/P_{\leq 0}.$$

We claim that this is the desired triangle. Indeed, $P/P_{\leq 0} \in D_{\mathcal{A}}^{f,\geq 1}$. Since $P_{\leq 0}$ has generators in negative degrees and is $\mathcal{K}$-projective (11.4.8), it lies in $D_{\mathcal{A}}^{f,\leq 0}$. This proves the theorem.

11.4.12. Recall that a t-structure on a triangulated category D defines the **truncation functors**

$$\tau_{\leq n} : D \to D^{\leq n}$$

$$\tau_{\geq n} : D \to D^{\geq n},$$

which are respectively right and left adjoint to the inclusions $D^{\leq n} \subset D, D^{\geq n} \subset D$. Then for $X \in D$ the exact triangle

$$\tau_{\leq 0} X \to X \to \tau_{\geq 1} X$$

is the unique triangle (up to a unique isomorphism) satisfying condition (iii) of definition 11.4.1 (see [BBD]).

In our case the truncation functors are made explicit by the argument in 11.4.11.above. Namely if P is a minimal $\mathcal{K}$-projective then $\tau_{\leq i} P = P_{\leq i}$ and $\tau_{\geq i+1} P = P/P_{\leq i}$.

11.4.13. Given a t-structure on D, its **heart** is the full subcategory $\mathcal{C} := D^{\geq 0} \cap D^{\leq 0}$. It is known ([BBD]) that $\mathcal{C}$ is **abelian**.

Claim. *The abelian category $D_{\mathcal{A}}^{f,\geq 0} \cap D_{\mathcal{A}}^{f,\leq 0}$ is equivalent to $\mathrm{Vect}_{\mathbb{R}}$ – the category of finite dimensional vector spaces over $\mathbb{R}$.*

Proof. Let $P \in D_{\mathcal{A}}^{f,\geq 0} \cap D_{\mathcal{A}}^{f,\leq 0}$. We may assume that P is minimal $\mathcal{K}$-projective. Since $P \in D_{\mathcal{A}}^{f,\leq 0}$, $\tau_{\geq 1} P = 0$. Since $P \in D_{\mathcal{A}}^{f,\geq 0}$, $\tau_{\leq -1} X = 0$. So by 11.4.12 we find that $P = \oplus A$, $d_P = 0$. Hence $D_{\mathcal{A}}^{f,\geq 0} \cap D_{\mathcal{A}}^{f,\leq 0}$ is equivalent to the category of free A-modules of finite rank, placed in degree zero, which in turn is equivalent to $\mathrm{Vect}_{\mathbb{R}}$.

11.4.14. One can characterize the subcategories $D_{\mathcal{A}}^{f,\geq 0}, D_{\mathcal{A}}^{f,\leq 0} \subset D_{\mathcal{A}}^{f}$ in the following way.

Proposition. *Let $M \in D_{\mathcal{A}}^{f}$.*

1. The following conditions are equivalent
(i) $M \in D_{\mathcal{A}}^{f,\geq 0}$.
(ii) There exists a $\mathcal{K}$-projective $P \in D_{\mathcal{A}}^{f}$ quasiisomorphic to M such that $P^i = 0, i < 0$.
(iii) If $P \in D_{\mathcal{A}}^{f}$ is a minimal $\mathcal{K}$-projective quasiisomorphic to M, then $P^i = 0, i < 0$.

2. The following conditions are equivalent

(iv) $M \in D_A^{f,\leq 0}$

(v) There exists a $\mathcal{K}$-projective $P \in D_A^f$ quasiisomorphic to M such that P is generated as an A-module by elements in nonpositive degrees.

(vi) If $P \in D_A^f$ is a minimal $\mathcal{K}$-projective quasiisomorphic to M, then P is generated as an A-module by elements in nonpositive degrees.

Proof. 1. Clearly, (iii) $\Rightarrow$ (ii) $\Rightarrow$ (i).

(i) $\Rightarrow$ (iii). Let $M \in D_A^{f,\geq 0}$ and let P be a minimal $\mathcal{K}$-projective quasiisomorphic to M. Since $P \in D_A^{f,\geq 0}$, $\tau_{\leq -1}P = 0$. But then by 11.4.12, $P^i = 0, i < 0$ which proves (iii).

2. Clearly, (vi) $\Rightarrow$ (v) $\Rightarrow$ (iv).

(iv) $\Rightarrow$ (vi). Let $M \in D_A^{f,\leq 0}$ and let P be a minimal $\mathcal{K}$-projective quasiisomorphic to M. Then $\tau_{\geq 1}P = 0$, so by 11.4.12 $P = \tau_{\leq 0}P = P_{\leq 0}$, which proves (vi).

12. DG–modules and sheaves on topological spaces.

This is a fairly technical section whose only purpose is to prove the main theorem 12.7.2. Otherwise, it is never used later.

12.0 In this section we show how DG–modules are connected with sheaves on topological spaces. Namely, to a topological space X one can associate a canonical DG–algebra $\mathcal{A}_X$, so that a continuous map $X \xrightarrow{f} Y$ defines a homomorphism of DG–algebras $\phi : \mathcal{A}_Y \to \mathcal{A}_X$. Let $D_{\mathcal{A}_X}$ be the derived category of left $\mathcal{A}_X$-modules and $D(X)$ be the derived category of sheaves on X. We define the localization functor

$$\mathcal{L}_X : D_{\mathcal{A}_X} \to D(X)$$

and the global sections functor

$$\gamma_X : D^+(X) \to D_{\mathcal{A}_X}.$$

These functors establish an equivalence between certain natural subcategories of $D_{\mathcal{A}_X}$ and $D(X)$. Then we study the compatibility of the localization functor with the inverse image $f^* : D(Y) \to D(X)$ and the direct image $f_* : D^+(X) \to D^+(Y)$. These results will be applied to the derived category of equivariant sheaves $D_G(pt)$.

12.1. DG–algebras associated to a topological space.

Let X be a topological space, C_X – the constant sheaf of R-modules on X (later on we will stick to the reals $R = \mathbb{R}$).

Definition. Let $0 \to C_X \to \mathcal{F}^{\cdot}$ be a resolution of the constant sheaf. We say that it is **multiplicative** if there is given a map of complexes $m : \mathcal{F}^{\cdot} \otimes \mathcal{F}^{\cdot} \to \mathcal{F}^{\cdot}$ which is associative and induces the ordinary multiplication on the subsheaf C_X. The resolution $\mathcal{F}^{\cdot}$ is called **acyclic** if all sheaves $\mathcal{F}^n$ are acyclic, i.e. $H^i(X, \mathcal{F}^n) = 0$, $i > 0$.

Given a multiplicative resolution $C_X \to \mathcal{F}^{\cdot}$ the complex of global sections $\Gamma(\mathcal{F}^{\cdot})$ has a structure of a DG–algebra. This algebra makes sense if $\mathcal{F}^{\cdot}$ is in addition acyclic; then, for example, $H^i(\Gamma(\mathcal{F}^{\cdot})) = H^i(X, C_X)$.

12.1.1. Examples. 1. The canonical Godement resolution $C_X \to C^{\cdot}$ (see [Go], 4.3).

2. The canonical simplicial Godement resolution $C_X \to \mathcal{F}^{\cdot}$ (see [Go], 6.4).
3. The resolution by localized singular cochains $C_X \to CS^{\cdot}$ (see [Go], 3.9).
4. If X is a manifold, one can take the resolution by the de Rham complex of smooth forms $C_X \to \Omega_X^{\cdot}$.

All resolutions in above examples are acyclic at least if X is paracompact. Notice that the first three resolutions are functorial with respect to continuous maps. Namely, given a continuous map $f : X \to Y$, we have a natural map $f^* B_Y^{\cdot} \to B_X^{\cdot}$

(where $B^{\cdot}$ is a resolution from examples 1-3) which induces the homomorphism of DG–algebras $\phi : \Gamma(B_Y) \to \Gamma(B_X)$. The de Rham complex $\Omega_X^{\cdot}$ is functorial with respect to smooth maps.

The next proposition shows that the choice of a particular acyclic resolution is not important.

12.1.2. Proposition. *Let X be a topological space and $C_X \to B^{\cdot}$ be an acyclic multiplicative resolution. Then the DG–algebra $\Gamma(B^{\cdot})$ is canonically quasiisomorphic to the DG–algebra $\Gamma(\mathcal{F}^{\cdot})$, where $C_X \to \mathcal{F}^{\cdot}$ is the simplicial Godement resolution. More precisely, there exists an acyclic multiplicative resolution $C_X \to \mathcal{F}^{\cdot}(B^{\cdot})$ and canonical morphisms $\mathcal{F}^{\cdot} \to \mathcal{F}^{\cdot}(B^{\cdot})$ and $B^{\cdot} \to \mathcal{F}^{\cdot}(B^{\cdot})$, which induce quasiisomorphisms on DG–algebras of global sections. In particular, any two DG–algebras coming from acyclic multiplicative resolutions of C_X are canonically quasiisomorphic, and hence the corresponding derived categories of DG–modules are canonically equivalent (10.12.5.1).*

Proof. Let us recall the simplicial Godement resolution $\mathcal{F}^{\cdot}$ (see [Go], 6.4). Let $A \in Sh(X)$. There exists a canonical resolution $\mathcal{F}^{\cdot}(A)$ of A:

$$0 \to A \to \mathcal{F}^0(A) \xrightarrow{d} \mathcal{F}^1(A) \xrightarrow{d} \dots,$$

where $\mathcal{F}^0(A) = C^0(A)$ – the sheaf of discontinuous sections of A (see [Go], 4.3) and $\mathcal{F}^n(A) = C^0(\mathcal{F}^{n-1}(A))$. We denote the resolution $\mathcal{F}^{\cdot}(C_X)$ simply by $\mathcal{F}^{\cdot}$.

Recall that local sections $s^n \in \mathcal{F}^n(A)(U)$ are represented by functions

$$s^n(x_0, \dots, x_n) \in A_{x_n}$$

defined on U^{n+1}. Two such functions define the same section if they satisfy certain equivalence relation (see [Go], 6.4).

Following Godement we will use the following convention. Let $u \in A_x$. Then we denote by $y \mapsto u(y) \in A_y$ any local (continuous) section of A which is equal to u when $y = x$. Using these notations we can write the differential

$$d : \mathcal{F}^n(A) \to \mathcal{F}^{n+1}(A)$$

as follows

$$(ds^n)(x_0, \dots, x_{n+1}) = \sum_{i=0}^{n} (-1)^i s^n(x_0, \dots, \hat{x}_i, \dots, x_{n+1})$$

$$+ (-1)^{n+1} s^n(x_0, \dots, x_n)(x_{n+1}).$$

The functor $A \mapsto \mathcal{F}^{\cdot}(A)$ has the following properties.

(1) It is exact.

(2) Each sheaf $\mathcal{F}^n(A)$ is flabby, hence acyclic.

(3) If A is a sheaf of rings, then $\mathcal{F}(A)$ has a multiplicative structure $\times$ defined by the formula

$$s^p \times s^q(x_0, \ldots, x_{p+q}) = s^p(x_0, \ldots, x_p)(x_{p+q}) \circ s^q(x_p, \ldots, x_{p+q}),$$

where $s^p \in \mathcal{F}^p(A), s^q \in \mathcal{F}^q(A)$ and $\circ$ denotes the multiplication in A.

Hence in particular $\mathcal{F}$ is an acyclic multiplicative resolution of C_X.

Let $C_X \to B^{\cdot}$ be an acyclic multiplicative resolution. Consider the double complex

$$
\begin{array}{ccccccc}
0 & \to & \mathcal{F}^1 & \to & \cdots & & \cdots \\
& & \uparrow & & \uparrow & & \\
0 & \to & \mathcal{F}^0 & \to & \mathcal{F}^0(B^0) & \to & \mathcal{F}^0(B^1) & \cdots \\
& & \uparrow & & \uparrow & & \uparrow \\
0 & \to & C_X & \to & B^0 & \to & B^1 & \to & \cdots \\
& & \uparrow & & \uparrow & & \uparrow \\
& & 0 & & 0 & & 0 &
\end{array}
$$

Denote by $\mathcal{F}(B^{\cdot})$ the total complex of the inside part

$$
\begin{array}{ccccc}
& & \cdots & & \\
& & \uparrow & & \\
0 & \to & \mathcal{F}^1(B^0) & \to & \cdots \\
& & \uparrow & & \uparrow \\
0 & \to & \mathcal{F}^0(B^0) & \to & \mathcal{F}^0(B^1) & \cdots \\
& & \uparrow & & \uparrow \\
& & 0 & & 0
\end{array}
$$

The complexes $\mathcal{F}$ and $B^{\cdot}$ embed naturally in $\mathcal{F}(B^{\cdot})$, and properties (1), (2) above imply that these embeddings induce quasiisomorphisms between the global sections.

So it remains to construct a multiplicative structure on $\mathcal{F}(B^{\cdot})$ which will agree with the given ones on $\mathcal{F}$ and $B^{\cdot}$. Let $\circ : B^{\cdot} \otimes B^{\cdot} \to B^{\cdot}$ denote the given multiplication on $B^{\cdot}$. Let us define the multiplication $\times : \mathcal{F}(B^{\cdot}) \otimes \mathcal{F}(B^{\cdot}) \to \mathcal{F}(B^{\cdot})$ as follows. Given $s^{p,i} \in \mathcal{F}^p(B^i), s^{q,j} \in \mathcal{F}^q(B^j)$, the section $s^{p,i} \times s^{q,j} \in \mathcal{F}^{p+q}(B^{i+j})$ is defined by the formula

$$s^{p,i} \times s^{q,j}(x_0, \ldots, x_{p+q}) = (-1)^{qi} s^{p,i}(x_0, \ldots, x_p)(x_{p+q}) \circ s^{q,j}(x_p, \ldots, x_{p+q}).$$

One checks immediately that $\times$ is a morphism of complexes and that it induces the given multiplication on $\mathcal{F}$ and $B^{\cdot}$. This proves the proposition.

As was mentioned above the de Rham algebra is functorial only with respect to smooth maps. However, it has the advantage of being supercommutative, and hence its category of DG–modules has more structure. Since in this work we are interested only in very special topological spaces – the classifying spaces for Lie groups – we will stick to the de Rham algebra.

12.2. The de Rham complex of an ∞–dimensional manifold.

For the remaining part of this section 12 let us put $R = \mathbb{R}$.

12.2.1. Definition. An ∞–dimensional manifold M is a paracompact topological space with a fixed homeomorphism,

$$M \simeq \varinjlim M_n,$$

where $M_1 \overset{j}{\hookrightarrow} M_2 \overset{j}{\hookrightarrow} M_3 \ldots$ is a sequence of smooth (paracompact) manifolds of increasing dimensions $d_1 < d_2 < \ldots$, and j is an embedding of a closed submanifold. (A subset $U \subset \varinjlim M_n$ is open iff $U \cap M_n$ is open in M_n for each n).

Let $\Omega_{M_n} := 0 \to \Omega^0_{M_n} \to \Omega^1_{M_n} \to \cdots \to \Omega^{d_n}_{M_n} \to 0$ be the de Rham complex of smooth differential forms on M_n. It is known (Poincare lemma), that Ω_{M_n} is a resolution of the constant sheaf C_{M_n}.

Extend the complex Ω_{M_n} by zero to M via the closed embedding $M_n \hookrightarrow M$ and denote this extension again by Ω_{M_n}. Then the restriction of forms from M_{n+1} to M_n produces the inverse system of complexes on M:

$$\cdots \to \Omega_{M_2} \to \Omega_{M_1}.$$

12.2.2. Definition. The **de Rham complex** on M is the inverse limit

$$\Omega_M := \varprojlim \Omega_{M_n}$$

12.2.3. Proposition. *(i) The complex Ω_M is a resolution of the constant sheaf C_M.*

(ii) Each sheaf $\Omega^k_M = \varprojlim \Omega^k_{M_n}$ is soft. Since M is paracompact, it follows that Ω^k_M is acyclic, i.e. $H^i(M, \Omega^k_M) = 0$, $i > 0$ (see [Go], 3.5.4).

Proof. (i) Fix a point $x \in M$, say $x \in M_n$. Let us show that the sequence of stalks $0 \to C_{M,x} \to \Omega^0_{M,x} \to \Omega^1_{M,x} \to \cdots$ is exact. It suffices to show that for a small open subset $U \ni x$ the complex of global sections

$$(*) \qquad 0 \to \Gamma(U, C_M) \to \Gamma(U, \Omega^0_M) \to \cdots$$

is exact. There is a fundamental system of neighborhoods of x consisting of open subsets U s.t. $U \cap M_k \simeq \mathbb{R}^{d_k}$, $k \geq n$, and $U \cap M_k \hookrightarrow U \cap M_{k+1}$ is the embedding of a plane $\mathbb{R}^{d_k} \hookrightarrow \mathbb{R}^{d_{k+1}}$. Then by the Poincare lemma the complex

$$0 \to \Gamma(U, C_{M_k}) \to \Gamma(U, \Omega^0_{M_k}) \to \cdots$$

is exact. Now the exactness of (*) follows since in the inverse system

$$\Gamma(U, \Omega_M^s) = \varprojlim_k \Gamma(U, \Omega_{M_k}^s)$$

all maps

$$\Gamma(U, \Omega_{M_{k+1}}^s) \to \Gamma(U, \Omega_{M_k}^s)$$

are surjective.

(ii) Since Ω_M^s is a module over the sheaf of ring Ω_M^0 ("smooth functions" on M), it suffices to prove the following

12.2.4. Lemma. *The sheaf Ω_M^0 is soft.*

Indeed, the lemma implies that Ω_M^s is soft (as a module over a soft sheaf of rings) (see [Go],3.7.1), and since M is paracompact, it is acyclic ([Go],3.5.4).

Proof of lemma. Since M is paracompact it is enough to prove the following statement. Each point $x \in M$ has a neighborhood U such that for any disjoint closed subsets $S, T \subset M$ which are contained in U there is $f \in \Omega_M^0(U)$ s.t. $f \equiv 1$ in some neighborhood of S and $\equiv 0$ in some neighborhood of T.

Fix $x \in M$. Let us choose $U \ni x$ such that $U_k := U \cap M_k \simeq \mathbb{R}^{d_k}$ is relatively compact in M_k and $U_k \hookrightarrow U_{k+1}$ is the embedding of the plane $\mathbb{R}^{d_k} \subseteq \mathbb{R}^{d_{k+1}}$. Let $S, T \subset M$ be disjoint closed subsets contained in U. Then the intersections $S_k := S \cap U_k$, $T_k := T \cap U_k$ are *compact* subsets in U_k. It is shown in [Go], 3.7 that there exists a smooth function f_k on U_k such that $f_k \equiv 1$ is a neighborhood of S_k and $f_k \equiv 0$ in a neighborhood of T_k. So it remains to choose f_{k+1} on U_{k+1} so that $f_{k+1}|_{U_k} = f_k$.

Suppose f_k is chosen. Denote again by f_k its extension to U_{k+1} using the product structure $U_{k+1} = U_k \times \mathbb{R}^{d_{k+1}-d_k}$. Let $\tilde{f}_{k+1}$ be a smooth function on U_{k+1} such that $\tilde{f}_{k+1} \equiv 1$ near S_{k+1} and $\tilde{f}_{k+1} \equiv 0$ near T_{k+1}. Since S_{k+1} and T_{k+1} are compact we can choose a small open neighborhood V of U_k in U_{k+1} with the following property. Put $W = U_{k+1} \backslash U_k$, and let φ_V, φ_W be a parition of 1 subject to the covering $U_{k+1} = V \cup W$. Then the function

$$f_{k+1} := \varphi_V \cdot f_k + \varphi_W \cdot \tilde{f}_{k+1}$$

will be equal to 1 in a neighborhood of S_{k+1} and equal to 0 in a neighborhood of T_{k+1}. Clearly $f_{k+1}|_{U_k} = f_k$ which proves the lemma and the proposition.

12.2.5. Since the restriction of forms $\Omega_{M_{n+1}} \to \Omega_{M_n}$ commutes with the wedge product, the de Rham complex Ω_M inherits a natural multiplicative structure. By the above proposition, $\Omega_M^{\cdot}$ is an acyclic multiplicative resolution of C_M (see definition 12.1). Denote by $\mathcal{A}_M$ the corresponding DG–algebra of global sections

$$\mathcal{A}_M := \Gamma(\Omega_M^{\cdot}).$$

12.2.6. Definition. Let $M' = \varinjlim M_n'$ be another ∞–dimensional manifold and $f : M \to M'$ be a continuous map. We say that f is **smooth** if for each n there exists n' such that $f(M_n) \subset M_{n'}'$, and the restriction

$$f|_{M_n} : M_n \to M_{n'}',$$

is smooth.

Let $f : M \to M'$ be a smooth map. Then we have a natural morphism $f^*\Omega_M^{\cdot} \to \Omega_{M'}^{\cdot}$ which preserves the product structure and hence defines the homomorphism of DG–algebras

$$\phi : \mathcal{A}_{M'} \to \mathcal{A}_M.$$

12.3. Localization and global sections.

In this section and in sections 12.4-6 below all spaces $X, Y, \ldots$ are smooth paracompact manifolds (possibly ∞–dimensional) and all maps $f : X \to Y$ are smooth. For a space X, $\mathcal{A}_X$ denotes the de Rham DG–algebra defined in 12.2.5.

12.3.1. Let us define the localization functor

$$\mathcal{L}_X : D_{\mathcal{A}_X} \to D(X),$$

where $D_{\mathcal{A}_X}$ is the derived category of (left) DG–modules over $\mathcal{A}_X$ (10.4.1) and $D(X)$ is the derived category of sheaves on X. Let $M \in D_{\mathcal{A}_X}$. Put

$$\mathcal{L}_X(M) := \Omega_X^{\cdot} \overset{L}{\otimes}_{\mathcal{A}_X} M.$$

In other words, let $P \to M$ be a $\mathcal{K}$–projective resolution of M (10.12.1, 10.12.4.5).

Then

$$\mathcal{L}_X(M) = \Omega_X^{\cdot} \otimes_{\mathcal{A}_X} P$$

which is the sheaf of complexes (or the complex of sheaves) on X associated to the presheaf of complexes

$$U \mapsto \Omega_X^{\cdot}(U) \otimes_{\mathcal{A}_X} P.$$

Note that $\mathcal{L}_X(\mathcal{A}_X) \simeq C_X$. Indeed, $\mathcal{A}_X$ is $\mathcal{K}$–projective as an $\mathcal{A}_X$–module (10.12.2.3), hence

$$\mathcal{L}_X(\mathcal{A}_X) = \Omega_X^{\cdot} \otimes_{\mathcal{A}_X} \mathcal{A}_X = \Omega_X^{\cdot} \simeq C_X.$$

12.3.2. Definition. Let D be a triangulated category, $S \in D$.

(i) Denote by $D(S) \subset D$ the full triangulated subcategory generated by S.

(ii) Consider "bounded below" direct sums $\oplus_{i>\mu}(\oplus S[-i])$. As in 11.1.4 we denote them by $\oplus^+ S[-i]$. Denote by $D^+(\oplus S) \subset D$ the full triangulated category generated by sums $\oplus^+ S[-i]$.

12.3.2.1. Remark. Let $\mathcal{A}$ be the DG-algebra studied in section 11. Take $D = D_{\mathcal{A}}$ and $S = \mathcal{A}$. Then $D(\mathcal{A}) = D_{\mathcal{A}}^f$ and $D^+(\oplus \mathcal{A}) = D_{\mathcal{A}}^+$ (11.1.5).

12.3.3. Proposition. *The localization functor induces an equivalence of categories*

$$\mathcal{L}_X : D(\mathcal{A}_X) \xrightarrow{\sim} D(C_X)$$

and

$$\mathcal{L}_X : D^+(\oplus \mathcal{A}_X) \xrightarrow{\sim} D^+(\oplus C_X).$$

Proof. We have $\mathcal{L}_X(\mathcal{A}_X) = C_X$. So to prove the first assertion we only have to check that

$$\operatorname{Hom}_{D_{\mathcal{A}_X}}(\mathcal{A}_X, \mathcal{A}_X[i]) = \operatorname{Hom}_{D(X)}(C_X, C_X[i]).$$

But

$$\operatorname{Hom}_{D_{\mathcal{A}}}(\mathcal{A}_X, \mathcal{A}_X[i]) = H^i(\mathcal{A}_X) = H^i(X, C_X) = \operatorname{Hom}_{D(x)}(C_X, C_X[i]).$$

Let us prove the second assertion. Let

$$M = \oplus^+ \mathcal{A}_X[-i], \ N = \oplus^+ \mathcal{A}_X[-j] \in D^+(\oplus \mathcal{A}_X).$$

Then

$$\mathcal{L}_X M = \oplus^+ C_X[-i], \ \mathcal{L}_X N = \oplus^+ C_X[-j] \in D^+(\oplus C_X),$$

since $\mathcal{L}_X$ preserves direct sums and the DG-modules M, N are $\mathcal{K}$-projective. It suffices to check that

$$\operatorname{Hom}_{D_{\mathcal{A}_X}}(M, N) = \operatorname{Hom}_{D(X)}(\mathcal{L}_X M, \mathcal{L}_X N).$$

Obviously, the left hand side is

$$\prod_i (\oplus_j H^{i-j}(\mathcal{A}_X)),$$

and the right hand side is

$$\prod_i (H^i(X, \oplus_j C_X[-i])).$$

So we need to show that

$$(*) \qquad H^i(X, \oplus_j C_X[-j]) = \oplus_j H^{i-j}(X, C_X).$$

Since X is a paracompact locally contractible space, we may use the singular cohomology to compute the groups in (*). Namely,

$$H^i(X, C_X) = H^i(X, \mathbb{R}),$$

and then the equality (*) follows from the universal coefficients formula. This proves the proposition.

The last argument also proves the following

12.3.4. Lemma. *Let* $\oplus^+ C_X[-i] \in D^+(\oplus C_X)$. *Consider its canonical soft resolution*

$$\oplus^+ C_X[-i] \to \oplus^+ \Omega_X^{\cdot}[-i].$$

Then the natural map of $\mathcal{A}_X$-modules

$$\oplus^+ \mathcal{A}_X[-i] \to \Gamma(X, \oplus^+ \Omega_X^{\cdot}[-i])$$

is a quasiisomorphism.

12.3.5. In order to define the functor of global sections

$$\gamma_X : D(X) \to D_{\mathcal{A}_X}$$

on the whole category $D(X)$ we need the notion of a $\mathcal{K}$-injective resolution (see [Sp]). To avoid the use of these resolutions and some other technical problems we prefer to work with the bounded below derived category $D^+(X)$. So we define the functor of global sections

$$\gamma_X : D^+(X) \to D_{\mathcal{A}_X}$$

as follows. Let $S^{\cdot} \in D^+(X)$ be a complex of sheaves. Then the complex of global sections of the tensor product $\Omega_X^{\cdot} \otimes_{C_X} S^{\cdot}$ has a natural structure of a left module over $\mathcal{A}_X = \Gamma(\Omega_X^{\cdot})$. Put

$$\gamma_X(S^{\cdot}) := \Gamma(\Omega_X^{\cdot} \otimes_{C_X} S^{\cdot}).$$

We must check that γ_X is well defined on $D^+(X)$, that is γ_X preserves quasiiso-morphisms. Note that the functor $\Omega_X^{\cdot} \otimes_{C_X} (\cdot)$ is exact, since we work with sheaves of $\mathbb{R}$–vector spaces. Also, the complex $\Omega_X^{\cdot} \otimes_{C_X} S^{\cdot}$ is bounded below and consists of sheaves $(\Omega_X^{\cdot} \otimes_{C_X} S^{\cdot})^m = \oplus_{k \geq 0} \Omega_X^k \otimes_{C_X} S^{m-k}$ which are modules over the soft sheaf of rings Ω_X^0 (12.2.4). Hence they are also soft and therefore acyclic for Γ, since X is paracompact ([Go],3.5.4). So γ_X is well defined.

Note that

$$\gamma_X(C_X) = \Gamma(\Omega'_X \otimes_{C_X} C_X) = \Gamma(\Omega'_X) = \mathcal{A}_X,$$

and by lemma 12.3.4

$$\gamma_X(\oplus^+ C_X[-i]) = \Gamma(\oplus^+ \Omega'_X[-i]) \simeq \oplus^+ \mathcal{A}_X[-i].$$

Hence γ_X maps subcategories $D(C_X), D^+(\oplus C_X) \subset D^+(X)$ to subcategories $D(\mathcal{A}_X)$ and $D^+(\mathcal{A}_X)$ respectively.

12.3.6. Proposition. *The functor*

$$\gamma_X : D^+(\oplus C_X) \to D^+(\oplus \mathcal{A}_X)$$

is an equivalence, which is the inverse to the equivalence

$$\mathcal{L}_X : D^+(\oplus \mathcal{A}_X) \to D^+(\oplus C_X)$$

of proposition 12.3.4. More precisely, there exist canonical isomorphisms of functors

$$\sigma : Id_{D^+(\oplus \mathcal{A}_X)} \to \gamma_X \cdot \mathcal{L}_X,$$

$$\tau : \mathcal{L}_X \cdot \gamma_X \to Id_{D^+(\oplus C_X)}.$$

Similarly, for $D(C_X)$ and $D(\mathcal{A}_X)$.

Proof. Let us define the morphism σ.

Let $M \in D^+(\oplus \mathcal{A}_X)$ be $\mathcal{K}$–projective. Since direct sums $\oplus^+ \mathcal{A}_X[-i]$ are $\mathcal{K}$-projective, we may (and will) assume that $M^i = 0$, $i << 0$. Then

$$\mathcal{L}_X(M) = \Omega'_X \otimes_{\mathcal{A}_X} M \in D^+(X),$$

and

$$\gamma_X \cdot \mathcal{L}_X(M) = \Gamma(\Omega_X \otimes_{C_X} (\Omega_X \otimes_{\mathcal{A}_X} M)).$$

Consider the map of complexes

$$\Omega'_X \otimes_{C_X} (\Omega'_X \otimes_{\mathcal{A}_X} M) \xrightarrow{m \otimes 1} \Omega'_X \otimes_{\mathcal{A}_X} M,$$

where $m : \Omega'_X \otimes \Omega'_X \to \Omega'_X$ is the multiplication. We claim that $m \otimes 1$ is a quasiisomorphism. This follows from the following lemma.

12.3.7. Lemma. *Let $S' \in D(X)$ be a complex of sheaves and*

$$t : \Omega'_X \otimes_{C_X} S' \to S'$$

be a morphism of complexes, such that $t(1 \otimes s) = s$. Then t is a quasiisomorphism.

Proof of lemma. Indeed, the inclusion

$$i : S^{\cdot} \to \Omega_X^{\cdot} \otimes_{C_X} S^{\cdot}, \ s \mapsto 1 \otimes s$$

is a quasiisomorphism, and $t \cdot i = id_S$.

Since the map $m \otimes 1$ is a quasiisomorphism of bounded below complexes consisting of soft (hence acyclic) sheaves, it induces the quasiisomorphism of left $\mathcal{A}_X$-modules

$$\alpha : \Gamma(\Omega_X \otimes_{C_X} (\Omega_X \otimes_{\mathcal{A}_X} M)) \xrightarrow{\alpha} \Gamma(\Omega_X^{\cdot} \otimes_{\mathcal{A}_X} M).$$

On the other hand there is the obvious morphism of left $\mathcal{A}_X$-modules

$$\beta : M \to \Gamma(\Omega_X^{\cdot} \otimes_{\mathcal{A}_X} M), \ m \mapsto 1 \otimes m.$$

Finally we define $\sigma = \alpha^{-1} \cdot \beta$.

Assume that $M = \oplus^+ \mathcal{A}_X[-i]$. Then β is a quasiisomorphism by lemma 12.3.4. Hence σ is a quasiisomorphism if $M \in D^+(\oplus \mathcal{A}_X)$.

Let us define the morphism τ.

Choose $S^{\cdot} \in D^+(\oplus C_X)$. Then $\gamma_X(S^{\cdot}) = \Gamma(\Omega_X^{\cdot} \otimes_{C_X} S^{\cdot})$. Choose a quasiisomorphism $P \to \gamma_X(S^{\cdot})$, where $P \in D_{\mathcal{A}_X}$ is a $\mathcal{K}$-projective $\mathcal{A}_X$-module. We have

$$\mathcal{L}_X \cdot \gamma_X(S^{\cdot}) = \mathcal{L}_X(P) = \Omega_X^{\cdot} \otimes_{\mathcal{A}_X} P$$

with the morphism

$$1 \otimes a : \Omega_X \otimes_{\mathcal{A}_X} P \to \Omega_X \otimes_{\mathcal{A}_X} \Gamma(\Omega_X^{\cdot} \otimes_{C_X} S^{\cdot}).$$

Compose it with the multiplication map

$$b : \Omega_X^{\cdot} \otimes_{\mathcal{A}_X} \Gamma(\Omega_X^{\cdot} \otimes_{C_X} S^{\cdot}) \to \Omega_X^{\cdot} \otimes_{C_X} S^{\cdot}$$

$$w \otimes w' \otimes s \mapsto ww' \otimes s.$$

to get the morphism

$$b \cdot (1 \otimes a) : \mathcal{L}_X \cdot \gamma_X(S^{\cdot}) \to \Omega_X^{\cdot} \otimes_{C_X} S^{\cdot}$$

On the other hand we the obvious quasiisomorphism

$$c : S^{\cdot} \to \Omega_X^{\cdot} \otimes_{C_X} S^{\cdot}$$

$$s \mapsto 1 \otimes s.$$

So we define the morphism τ as the composition

$$\tau = c^{-1} \cdot b \cdot (1 \otimes a) : \mathcal{L}_X \cdot \gamma_X \to Id.$$

Assume that $S^{\cdot} = \oplus^{+} C_X[-i]$. Then $\gamma_X(S^{\cdot})$ is quasiisomorphic to $\oplus^{+} \mathcal{A}_X[-i]$ (lemma 12.3.4), which is $\mathcal{K}$- projective, so we may take $P = \oplus^{+} \mathcal{A}_X[-i]$. Then the map

$$b \cdot (1 \otimes a) : \Omega_X^{\cdot} \otimes_{\mathcal{A}_X} (\oplus^{+} \mathcal{A}_X[-i]) \to \Omega_X^{\cdot} \otimes_{C_X} (\oplus^{+} C_X[-i])$$

is an isomorphism. Hence τ is a quasiisomorphism if $S^{\cdot} \in D^{+}(\oplus C_X)$. This proves the proposition.

12.3.8. Remark. All the results in this section 12.3 are valid for general paracompact locally contractible spaces X and a DG-algebra $\Gamma(X, \mathcal{F}^{\cdot})$ for a multiplicative acyclic resolution $\mathcal{F}^{\cdot}$ of C_X (the sheaf $\mathcal{F}^0$ must be soft and the basic ring R must be a field). In particular we never used the fact that $\mathcal{A}_X$ was supercommutative.

12.4. Applications to classifying spaces.

We want to apply the results of previous sections 12.1–12.3 to "smooth models" of classifying spaces.

Let G be a Lie group.

12.4.1. Defintion. A **smooth classifying sequence** for G is a sequence of closed embeddings

$$M_0 \subset M_1 \subset \ldots,$$

where M_k is a free k–acyclic smooth paracompact G–space, $M_k \subset M_{k+1}$ is an embedding of a submanifold and $\dim(M_{k+1}) > \dim(M_k)$.

Let $M_0 \subset M_1 \subset \ldots$ be a smooth classifying sequence for G. Denote the quotient $G \backslash M_k = BG_k$. Then we get a sequence of closed embeddings of smooth manifolds $BG_0 \subset BG_1 \subset \ldots$. The classifying space $BG = \lim_{\longrightarrow} BG_k$ is a smooth ∞–dimensional manifold (12.2.1). We call is a **smooth model** or a smooth classifying space.

12.4.2. Lemma. *Assume that the Lie group G has one of the following properties*
  (a) G is a linear group, i.e. a closed subgroup of $GL(n, \mathbb{R})$ for some n.
  (b) G has a finite number of connected components.
  Then there exists a smooth classifying sequence for G.

Proof. (a) Let M_k denote the Stiefel manifold of n–frames in $\mathbb{R}^{n+k}$. Then the sequence

$$M_0 \subset M_1 \subset \ldots$$

is a smooth classifying sequence for G.

  (b) Let $K \subset G$ be a maximal compact subgroup.

By a theorem of G. Mostow G/K is contractible. By the Peter–Weyl theorem K is linear and so by (a) there exists a smooth classifying sequence for K

$$M_0 \subset M_1 \subset \ldots .$$

Then

$$G \times_K M_0 \subset G \times_K M_1 \subset \ldots$$

is a smooth classifying sequence for G. This proves the lemma.

12.4.3. In the rest of this section 12.4 we will consider only **connected** Lie groups. Let G be such a group, and BG be its smooth classifying space (12.4.1). The derived category $D^b_{G,c}(pt)$ of G–equivariant constructible sheaves on pt is canonically equivalent to the full subcategory of $D^b(BG)$ consisting of complexes with constant cohomology sheaves of finite rank (see 2.7.2,2.8). This last category is generated by the constant sheaf C_{BG}. In other words

(1) $$D^b_{G,c}(pt) = D(C_{BG}).$$

But by (12.3.3) $D(\mathcal{A}_{BG}) \simeq D(C_{BG})$, where $\mathcal{A}_{BG}$ is the de Rham algebra of the smooth space BG. So we obtain an equivalence of triangulated categories

(2) $$D(\mathcal{A}_{BG}) \simeq D^b_{G,c}(pt)$$

We will go one step further and make the left hand side of (2) more accessible.

It is known that the cohomology ring $H^*(BG, \mathbb{R})$ is isomorphic to a polynomial ring $\mathbb{R}[X_1, \ldots, X_n]$, where generators X_i have various *even* degrees. Denote this ring by A_G and consider the DG–algebra

$$\mathcal{A}_G := (A_G, d = 0)$$

as in section 11.

12.4.4. Proposition. *There exists a homomorphism of DG–algebras $\mathcal{A}_G \to \mathcal{A}_{BG}$ which is a quasiisomorphism; hence it induces an equivalence of categories $D_{\mathcal{A}_G} \simeq D_{\mathcal{A}_{BG}}$ (10.12.5.1). This equivalence is unique up to a canonical isomorphism.*

Proof. Choose differential forms $\phi_1, \ldots, \phi_n \in \mathcal{A}_{BG}$ which represent cohomology classes $X_1, \ldots, X_n$. Since the degrees of X_i's are even the forms ϕ_i generate a *commutative* subalgebra in $\mathcal{A}_{BG}$. Hence we may define a homomorphism

$$\phi : \mathcal{A}_G \to \mathcal{A}_{BG} , \quad X_i \mapsto \phi_i ,$$

which is clearly a quasiisomorphism. It defines an equivalence of categories

$$\phi^* : D_{\mathcal{A}_G} \xrightarrow{\sim} D_{\mathcal{A}_{BG}}.$$

Let $\psi_1, \ldots, \psi_n \in \mathcal{A}_{BG}$ be a difference choice of forms which induces the corresponding equivalence

$$\psi^* : D_{\mathcal{A}_G} \xrightarrow{\sim} D_{\mathcal{A}_{BG}}.$$

We will show that functors ϕ^*, ψ^* are canonically isomorphic.

Since ϕ_i, ψ_i represent the same cohomology class we can choose $w_i \in \mathcal{A}_{BG}$ such that $dw_i = \phi_i - \psi_i$. Let $C \subset \mathcal{A}_{BG}$ be the DG–subalgebra generated by $\{\varphi_i, \psi_i, w_i\}$, and $\gamma : C \to \mathcal{A}_{BG}$ be the inclusion. We have two natural embeddings

$$\alpha, \beta : \mathcal{A}_G \to C,$$

where $\alpha(X_i) = \phi_i$, $\beta(X_i) = \psi_i$. Consider the induced equivalences of categories

$$\alpha^*, \beta^* : D_{\mathcal{A}_G} \to D_C, \quad \gamma^* : D_C \to D_{\mathcal{A}_{BG}}.$$

Since $\phi^* = \gamma^* \cdot \alpha^*$, $\psi^* = \gamma^* \cdot \beta^*$ it suffices to construct an isomorphism of functors $\alpha^* \xrightarrow{\sim} \beta^*$.

Let $\delta : C \to \mathcal{A}_G$ be the homomorphism $\delta(\phi_i) = \delta(\psi_i) = X_i$, $\delta(w_i) = 0$, and let $\delta^* : D_C \to D_{\mathcal{A}_G}$ be the corresponding equivalence. Note that $\delta^* \cdot \alpha^* = id = \delta^* \cdot \beta^*$. Hence functors α^*, β^* are canonically isomorphic.

To complete the proof of the proposition we must show that a different choice of forms w_i will produce the same isomorphism $\phi^* \xrightarrow{\sim} \psi^*$. We will only sketch the argument since it is similar to the one just given.

Let w_i' be a different choice of forms that produces the DG–subalgebra $C' \subset \mathcal{A}_{BG}$. Since $d(w_i - w_i') = 0$ and the odd cohomology of BG vanishes we can find η_i such that $d\eta_i = w_i - w_i'$. Let $E \subset \mathcal{A}_{BG}$ be the subalgebra generated by $\{\phi_i, \psi_i, w_i, w_i', \eta_i\}$. Now all algebras $\mathcal{A}_G, C, C'$ embed in E and it suffices to prove the equality of two morphisms in D_E, which is done similarly. This proves the proposition.

12.4.5. Composing the equivalence of 12.4.4 with the localization functor of 12.4.3 we obtain the functor

$$\mathcal{L}_G := D_{\mathcal{A}_G} \to D(BG)$$

which induces an equivalence

$$\mathcal{L}_G : D(\mathcal{A}_G) \xrightarrow{\sim} D_{G,c}^b(pt).$$

But $D(\mathcal{A}_G)$ is the category $D_{\mathcal{A}_G}^f$ studied in detail in section 11 (see 11.1.5). We call the obtained equivalence

$$\mathcal{L}_G : D_{\mathcal{A}_G}^f \xrightarrow{\sim} D_{G,c}^b(pt)$$

the localization.

12.4.6. Proposition. *The localization functor*

$$\mathcal{L}_G : D_{\mathcal{A}_G}^f \xrightarrow{\sim} D_{G,c}^b(pt)$$

is an equivalence of t-categories, which commutes with functors $\overset{L}{\otimes}$, $RHom$, D and the cohomological functor $H : D^f_{\mathcal{A}_G} \to Mod_{\mathcal{A}_G}$.

Proof. Let $P \in D^f_{\mathcal{A}_G}$ be a minimal $\mathcal{K}$-projective (11.4.6). Then $P \in D^{f,\geq 0}_{\mathcal{A}_G} \Leftrightarrow P^i = 0$, $i < 0$ and $P \in D^{f,\leq 0}_{\mathcal{A}_G} \Leftrightarrow P$ is generated by elements in nonpositive degrees (see prop. 11.4.14). Note that $\mathcal{L}_G(\mathcal{A}_G) = C_{BG}$. Hence $\mathcal{L}_G$ preserves the subcategories $D^{\geq 0}$ and $D^{\leq 0}$ and so is an equivalence of t-categories.

Let $M, N \in D^f_{\mathcal{A}_G}$ be two $\mathcal{K}$-projective DG–modules. Then

$$\mathcal{L}_G(M \overset{L}{\otimes} N) = \Omega_{BG} \otimes_{\mathcal{A}_G} (M \otimes_{\mathcal{A}_G} N).$$

Define a morphism of complexes

$$\theta : \mathcal{L}_G(M) \otimes_{C_{B_G}} \mathcal{L}_G(N) \to \mathcal{L}_G(M \overset{L}{\otimes} N)$$

by the formula

$$\theta : (w \otimes m) \otimes (w' \otimes n) \mapsto (-1)^{\deg(m)\deg(w')} ww' \otimes m \otimes n.$$

Since θ is a quasiisomorphism if $M = N = \mathcal{A}_G$, it is an isomorphism of functors. Hence $\mathcal{L}_G$ commutes with $\overset{L}{\otimes}$.

To prove the statement for $RHom$ we need the following.

12.4.7. Lemma. *Let* $M, N \in D^f_{\mathcal{A}_G}$ *be* $\mathcal{K}$-*projective. Then* $Hom\ (M, N)$ *is also so.*

Proof. Since the subcategory of $\mathcal{K}$-projectives in $\mathcal{K}^f_{\mathcal{A}_G}$ is generated by $\mathcal{A}_G$, it suffices to prove the lemma for $M = \mathcal{A}_G[i], N = \mathcal{A}_G[j]$, in which case it is obvious.

Let $M, N \in D^f_{\mathcal{A}_G}$ be $\mathcal{K}$-projective. Then by the lemma

$$\mathcal{L}_G(RHom(M, N)) = \Omega_{BG} \otimes_{\mathcal{A}_G} Hom(M, N).$$

Let $i : \mathcal{L}_G(N) \to I$ be an injective resolution. Then

$$RHom(\mathcal{L}_G(M), \mathcal{L}_G(N)) = Hom(\Omega_{BG} \otimes_{\mathcal{A}_G} M, I).$$

Define a morphism of complexes

$$\delta : \mathcal{L}_G(RHom(M, N)) \to RHom(\mathcal{L}_G(M)\mathcal{L}_G(N))$$

by the formula

$$\delta(w \otimes f)(w' \otimes m) = (-1)^{\deg(f)\deg(w')} i(ww' \otimes f(m)).$$

It is a quasiisomorphism if $M = N = \mathcal{A}_G$, hence is an isomorphism of functors. So $\mathcal{L}_G$ commutes with $RHom$. Since $\mathcal{L}_G(\mathcal{A}_G) = C_{BG}$, it also commutes with the duality D. It remains to treat the cohomological functor H.

Let $M \in D^f_{\mathcal{A}_G}$ be $\mathcal{K}$-projective. We have a map of DG–modules

$$\gamma : M \to \Gamma(\Omega_{BG} \otimes_{\mathcal{A}_G} M) = \Gamma(\mathcal{L}_G(M))$$

$$m \mapsto 1 \otimes m.$$

It is a quasiisomorphism if $M = \mathcal{A}_G$, hence is so in general. This proves the proposition.

12.4.8. Proposition. *The localization functor* $\mathcal{L}_G : D_{\mathcal{A}_G} \to D(BG)$ *induces an equivalence of full subcategories*

$$\mathcal{L}_G : D^+_{\mathcal{A}_G} \xrightarrow{\sim} D^+_G(pt)$$

(see 11.1,11.1.5). It commutes with $\overset{L}{\otimes}$ *and* H.

Proof. We know that $D^+_{\mathcal{A}_G}$ is generated by bounded below direct sums $\oplus^+ \mathcal{A}_G[-i]$ (11.1.5). If $M = \oplus^+ \mathcal{A}_G[i], N = \oplus^+ \mathcal{A}_G[j] \in D^+_{\mathcal{A}_G}$ are two such modules then

$$\mathrm{Hom}_{D_{\mathcal{A}_G}}(M, N) = \mathrm{Hom}_{D(BG)}(\mathcal{L}_G(M), \mathcal{L}_G(N)),$$

by proposition 12.3.3, so $\mathcal{L}_G$ is an equivalence of $D^+_{\mathcal{A}_G}$ with its essential image in $D(BG)$. Since $\mathcal{L}_G(D^+_{\mathcal{A}_G}) \subset D^+_G(pt)$ it remains to show that any complex $S \in D^+(BG)$ with constant cohomology sheaves lies in $\mathcal{L}_G(D^+_{\mathcal{A}_G})$.

Let S be such a complex. Then S is the direct limit

$$S = \lim_{\to} \tau_{\leq n} S.$$

Each $\tau_{\leq n} S$ lies in $\mathcal{L}_G(D^+_{\mathcal{A}_G})$, say $\tau_{\leq n} S \simeq \mathcal{L}_G(M_n)$ for a $\mathcal{K}$-projective M_n. The modules M_n form a corresponding direct system in $D^+_{\mathcal{A}_G}$ and if we put

$$M := \lim_{\to} M_n \in D_{\mathcal{A}_G},$$

then $\mathcal{L}_G(M) = S$, since $\mathcal{L}_G$ preserves direct limits. It only remains to show that M lies in $D^+_{\mathcal{A}_G}$.

Note that $\tau_{\leq n+1} S = C(H^{n+1}(S)[-1] \to \tau_{\leq n} S)$. Hence we can put M_{n+1} to be

$$M_{n+1} = C(H^{n+1}(S)[-1] \otimes_R \mathcal{A}_G \to M_n),$$

so that if $M^i_n = 0$ for $i < m$, then the same is true for M_{n+1}. But then clearly

$$M = \lim_{\to} M_n = \cup_n M_n \in D^+_{\mathcal{A}_G}$$

This proves the first part of the proposition.

The second part is prove similarly to proposition 12.4.6 above using lemma 12.3.4.

12.4.9. Corollary. *The subcategory $D_G^+(pt) \subset D(BG)$ coincides with $D^+(\oplus C_{BG})$.*

Proof. This follows from 11.1.5, 12.3.3, 12.4.8.

12.4.10. To conclude this section we want to show that the constructed equivalences

$$\mathcal{L}_G : D_{\mathcal{A}_G}^f \rightarrow D_{G,c}^b(pt)$$

$$\mathcal{L}_G : D_{\mathcal{A}_G}^+ \rightarrow D_G^+(pt)$$

do not depend on the choice of a smooth model for BG.

Let

$$M_1 \subset M_2 \subset \ldots$$

$$M_1' \subset M_2' \subset \ldots$$

be two smooth classifying sequences for G giving rise to smooth models BG, BG'. Consider the product sequence

$$M_1 \times M_1' \subset M_2 \times M_2' \subset \ldots$$

which produces the smooth model BG''. We have the diagram of smooth maps

$$
\begin{array}{ccc}
 & BG'' & \\
p \swarrow & & \searrow p' \\
BG & & BG'
\end{array}
$$

and the corresponding homomorphisms of DG–algebras

$$
\begin{array}{ccc}
 & \mathcal{A}_{BG''} & \\
\phi \nearrow & & \phi' \nwarrow \\
\mathcal{A}_{BG} & & \mathcal{A}_{BG'}
\end{array}
$$

So we may assume that $BG' = BG''$. Then the required result follows from the commutativity of the functorial diagram

$$
\begin{array}{ccc}
D_{\mathcal{A}_{BG''}} & \overset{\mathcal{L}}{\rightarrow} & D(BG'') \\
\phi^* \uparrow & & \uparrow p^* \\
D_{\mathcal{A}_{BG}} & \overset{\mathcal{L}}{\rightarrow} & D(BG),
\end{array}
$$

which is a special case of proposition 12.5.1 below.

12.5. Localization and inverse image.

We keep the notations of section 12.3. Let $f : X \to Y$ be a (smooth) map. It induces functors of inverse and direct image $f^* : D(Y) \to D(X), f_* : D(X) \to D(Y)$ (see [Sp]). On the other hand we have the corresponding homomorphism of DG-algebras $\psi : \mathcal{A}_Y \to \mathcal{A}_X$ which induces functors $\psi^* : D_{\mathcal{A}_Y} \to D_{\mathcal{A}_X}, \psi_* : \mathcal{A}_X \to D_{\mathcal{A}_Y}$. It is natural to ask if the above functors are compatible with the localization (see 12.3), i.e. if the following diagrams are commutative.

(1)
$$
\begin{array}{ccc}
D_{\mathcal{A}_X} & \overset{\mathcal{L}_X}{\rightrightarrows} & D(X) \\
\psi^* \uparrow & & \uparrow f^* \\
D_{\mathcal{A}_Y} & \overset{\mathcal{L}_Y}{\rightrightarrows} & D(Y)
\end{array}
$$

(2)
$$
\begin{array}{ccc}
D_{\mathcal{A}_X} & \overset{\mathcal{L}_X}{\rightrightarrows} & D(X) \\
\psi_* \downarrow & & \downarrow f_* \\
D_{\mathcal{A}_Y} & \overset{\mathcal{L}_Y}{\rightrightarrows} & D(Y)
\end{array}
$$

Here we discuss the inverse image. The direct image is discussed in the next section 12.6.

12.5.1. Proposition. *The diagram (1) is commutative. More precisely, there exists a canonical isomorphism of functors $f^* \cdot \mathcal{L}_Y \overset{\sim}{\to} \mathcal{L}_X \cdot \psi^*$.*

Proof. Let $N \in D_{\mathcal{A}_Y}$ be $\mathcal{K}$-projective. Then $\psi^*(N) = \mathcal{A}_X \otimes_{\mathcal{A}_Y} N \in D_{\mathcal{A}_X}$ is also $\mathcal{K}$-projective. So

$$
\mathcal{L}_X \cdot \psi^*(N) = \Omega_X \otimes_{\mathcal{A}_X} (\mathcal{A}_X \otimes_{\mathcal{A}_Y} N) = \Omega_X \otimes_{\mathcal{A}_Y} N.
$$

Also $\mathcal{L}_Y(N) = \Omega_Y \otimes_{\mathcal{A}_Y} N$ and

$$
f^* \cdot \mathcal{L}_Y(N) = f^*(\Omega_Y \otimes_{\mathcal{A}_Y} N)
$$

Given open subsets $U \subset X, V \subset Y$ such that $f(U) \subset V$ we have the natural map of right $\mathcal{A}_Y$-modules $\Omega_Y(V) \to \Omega_X(U)$ which induces a quasiisomorphism on the stalks

$$
\Omega_{Y,f(x)} \overset{\sim}{\to} \Omega_{X,x}
$$

for each $x \in X$. We get the corresponding map of complexes

$$
f^*(\Omega_Y \otimes_{\mathcal{A}_Y} N) \to \Omega_X \otimes_{\mathcal{A}_Y} N
$$

which is also a quasiisomorphism on stalks since N is $\mathcal{K}$–projective. This proves the proposition.

12.6. Localization and the direct image.

We keep notations of sections 12.3, 12.5. Let $f : X \to Y$ be a map, and $\psi : \mathcal{A}_Y \to \mathcal{A}_X$ be the correspondong homomorphism of DG-algebras. Recall (12.3.3) the equivalences

$$\mathcal{L}_X : D^+(\oplus \mathcal{A}_X) \to D^+(\oplus C_X),$$

$$\mathcal{L}_Y : D^+(\oplus \mathcal{A}_Y) \to D^+(\oplus C_Y).$$

12.6.1. Proposition. *Assume that the direct image f_* maps $D^+(\oplus C_X)$ to $D^+(\oplus C_Y)$. Then ψ_* maps $D^+(\oplus \mathcal{A}_X)$ to $D^+(\oplus \mathcal{A}_Y)$ and the following functorial giagram is commutative*

$$
\begin{array}{ccc}
D^+(\oplus \mathcal{A}_X) & \overset{\mathcal{L}_X}{\to} & D^+(\oplus C_X) \\
\psi_* \downarrow & & \downarrow f_* \\
D^+(\oplus \mathcal{A}_Y) & \overset{\mathcal{L}_X}{\to} & D^+(\oplus C_Y)
\end{array}
$$

Proof. Recall that the functor

$$\mathcal{L}_Y : D^+(\oplus \mathcal{A}_Y) \to D^+(\oplus C_Y)$$

has the inverse

$$\gamma_Y : D^+(\oplus C_Y) \to D^+(\oplus \mathcal{A}_Y)$$

with canonical isomorphisms

$$\sigma : Id \to \gamma_Y \cdot \mathcal{L}_Y,$$

$$\tau : \mathcal{L}_Y \cdot \gamma_Y \to Id$$

(see 12.3.6). So it suffices to construct an isomorphism of functors

$$\alpha : \psi_* \to \gamma_Y \cdot f_* \cdot \mathcal{L}_X$$

from $D^+(\oplus \mathcal{A}_X)$ to $D^+(\oplus \mathcal{A}_Y)$.

Let $M \in D^+(\oplus \mathcal{A}_X)$ be $\mathcal{K}$-projective. We may (and will) assume that $M^i = 0$, $i << 0$. Then $\psi_* M = M$ considered as an $\mathcal{A}_Y$ -module and

$$\gamma_Y \cdot f_* \cdot \mathcal{L}_X(M) = \gamma_Y(f_*(\Omega_X \otimes_{\mathcal{A}_X} M))$$

$$= \Gamma(\Omega_Y \otimes_{C_Y} f_*(\Omega_X \otimes_{\mathcal{A}_X} M))$$

(Here we use the fact that $\Omega_X \otimes_{\mathcal{A}_X} M$ is bounded below and consists of soft sheaves on a paracompact space, hence acyclic for f_*).

The multiplication map

$$\Omega_Y \otimes f_* \Omega_X \to f_* \Omega_X$$

induces a quasiisomorphism of bounded below complexes of soft sheaves

$$\Omega_Y \otimes f_*(\Omega_X \otimes_{\mathcal{A}_X} M) \to f_*(\Omega_X \otimes_{\mathcal{A}_X} M)$$

(lemma 12.3.7) and hence a quasiisomorphism of left $\mathcal{A}_Y$-modules

$$a : \gamma_Y \cdot f_* \cdot \mathcal{L}_X(M) \xrightarrow{\sim} \Gamma(f_*(\Omega_X \otimes_{\mathcal{A}_X} M)$$

$$= \Gamma(\Omega_X \otimes_{\mathcal{A}_X} M).$$

On the other hand the canonical morphism of $\mathcal{A}_X$-modules

$$b : M \to \Gamma(\Omega_X \otimes_{\mathcal{A}_X} M) \,, \; m \mapsto 1 \otimes m$$

is a quasiisomorphism, since $M \in D^+(\oplus \mathcal{A}_X)$ (12.3.4). Hence we may put

$$\alpha = a^{-1} \cdot b.$$

This proves the proposition.

12.7. Applications to $D_G(pt)$.

2.7.0. Let $\phi : H \to G$ be a homomorphism of **connected** Lie groups. Let

$$M_0 \subset M_1 \subset \ldots$$

$$N_0 \subset N_1 \subset \ldots$$

be smooth classifying sequences for H and G respectively (12.4.1). Then

$$M_0 \times N_0 \subset M_1 \times N_1 \subset \ldots$$

is also a smooth classifying sequence for H and projections

$$M_i \times N_i \to N_i$$

induce a smooth map of the corresponding smooth models

$$f : BH \to BG$$

Recall that we have canonical identifications of $D^b_{H,c}(pt), D^+_H(pt)$ as certain full subcategories of $D^+(BH)$, and similar for G (2.7.2, 2.9.5). Consider

$$pt \to pt$$

as a ϕ-map. Then we get the functors

$$Q^* : D^b_{G,c}(pt) \to D^b_{H,c}(pt)$$

$$Q^* : D^+_G(pt) \to D^+_H(pt)$$

$$Q_* : D^+_H(pt) \to D^+_G(pt)$$

which under the above identification coincide with

$$f^* : D^+(BG) \to D^+(BH)$$

$$f_* : D^+(BH) \to D^+(BG)$$

respectively (6.11).Recall that we have the equivalences of categories

$$\mathcal{L}_H : D^+_{\mathcal{A}_H} \xrightarrow{\sim} D^+_H(pt)$$

$$\mathcal{L}_G : D^+_{\mathcal{A}_G} \xrightarrow{\sim} D^+_G(pt)$$

(12.4.8). The map f induces a homomorphism of the cohomology rings $A_G \to A_H$ and hence two functors $\psi^* : D^+_{\mathcal{A}_G} \to D^+_{\mathcal{A}_H}, \psi_* : D^+_{\mathcal{A}_H} \to D^+_{\mathcal{A}_G}$.

Consider the functional diagrams

$$
\begin{array}{ccc}
D^+_{\mathcal{A}_H} & \xrightarrow{\mathcal{L}_H} & D^+_H(pt) \\
\psi^* \uparrow & & \uparrow Q^* \\
D^+_{\mathcal{A}_G} & \xrightarrow{\mathcal{L}_G} & D^+_G(pt)
\end{array}
$$

(1)

and

$$
\begin{array}{ccc}
D^+_{\mathcal{A}_H} & \xrightarrow{\mathcal{L}_H} & D^+_H(pt) \\
\psi_* \downarrow & & \downarrow Q_* \\
D^+_{\mathcal{A}_G} & \xrightarrow{\mathcal{L}_G} & D^+_G(pt)
\end{array}
$$

(2)

12.7.1. Proposition. *The above diagrams (1) and (2) are commutative.*

Proof. Recall (12.4.5) that the localization functor $\mathcal{L}_H$ is the composition of the equivalence $D_{\mathcal{A}_H} \tilde{\to} D_{\mathcal{A}_{BH}}$ (12.4.4) with the localization $\mathcal{L}_{BH} : D_{\mathcal{A}_{BH}} \to D(BH)$ (and similar for G).

We have the obvious commutative diagrams

$$
\begin{array}{ccc}
D_{\mathcal{A}_H} & \tilde{\to} & D_{\mathcal{A}_{BH}} \\
\psi^* \uparrow & & \uparrow \psi^* \\
D_{\mathcal{A}_G} & \tilde{\to} & D_{\mathcal{A}_{BG}}
\end{array}
$$

and

$$
\begin{array}{ccc}
D_{\mathcal{A}_H} & \tilde{\to} & D_{\mathcal{A}_{BH}} \\
\psi_* \downarrow & & \downarrow \psi_* \\
D_{\mathcal{A}_G} & \tilde{\to} & D_{\mathcal{A}_{BG}}.
\end{array}
$$

Hence the commutativity of (1) follows from 12.5.1.

By the corollary 12.4.9 the category $D^+(pt)$ coincides with $D^+(\oplus C_{BH})$ (and similar for G). Hence f_* maps $D^+(\oplus C_{BH})$ to $D^+(\oplus C_{BG})$ and therefore by proposition 12.6.1 the following diagram is commutative

$$
\begin{array}{ccc}
D^+(\oplus \mathcal{A}_{BH}) & \xrightarrow{\mathcal{L}_{BH}} & D_H^+(pt) \\
\psi_* \downarrow & & \downarrow Q_* \\
D^+(\oplus \mathcal{A}_{BG}) & \xrightarrow{\mathcal{L}_{BG}} & D_G^+(pt)
\end{array}
$$

is commutative. But $D_{\mathcal{A}_H}^+ = D^+(\oplus \mathcal{A}_H)$ (11.1.5) and similar for G. Hence the diagram (2) is also commutative which proves the proposition.

Let us now summarize the results of 12.4.6, 12.4.8, 12.7.1 in the following

12.7.2. Main Theorem. *Let G be a connected Lie group, $A_G = H(BG)$. Let $\mathcal{A}_G = (A_G, d = 0)$ be the corresponding DG-algebra.*

(i) There exists an equivalence of triangulated categories

$$\mathcal{L}_G : D_{\mathcal{A}_G}^+ \tilde{\to} D_G^+(pt)$$

which is unique up to a canonical isomorphism. It commutes with $\overset{L}{\otimes}$ and the cohomology functor $(\cdot) \overset{H}{\to} Mod_{A_G}$.

(ii) The above equivalence restricts to the functor between the full subcategories

$$\mathcal{L}_G : D_{\mathcal{A}_G}^f \tilde{\to} D_{G,c}^b(pt),$$

which is an equivalence of t–categories commuting with $\overset{L}{\otimes}$, RHom, D and H.

(iii) Let $\phi : H \to G$ be a homomorphism of connected Lie groups. It induces a homorphism of rings $\psi : A_G \to A_H$ and hence the functors of extension and restriction of scalars

$$\psi^* : D^+_{A_G} \to D^+_{A_H}$$

$$\psi_* : D^+_{A_H} \to D^+_{A_G}.$$

Let $Q^* : D^+_G(pt) \to D^+_H(pt)$ and $Q_* : D^+_H(pt) \to D^+_G(pt)$ be the functors of the inverse and direct image corresponding to the ϕ–map

$$pt \to pt.$$

Then under the identification of (i) we have $\psi^* = Q^*, \psi_* = Q_*$.

The above theorem gives an algebraic interpretation of the category $D_G(pt)$ and is our main tool in applications.

13. Equivariant cohomology.

13.0. Let G be a Lie group. Let X be a G–space and $p : X \to pt$ be the map to a point. It induces the direct image functor

$$p_* : D_G^+(X) \to D_G^+(pt),$$

where the category $D_G^+(pt)$ can be naturally realized as a full subcategory of $D^+(BG)$ for the classifying space BG (2.9.5). Put $A_G = H(BG, R)$. Notice that for $S \in D(BG)$ its cohomology $H(S) = H(BG, S)$ is naturally a graded A_G–module.

13.1. Definition. Let $F \in D_G^+(X)$. The G–**equivariant cohomology** $H_G(X, F)$ of X with coefficients in F is by definition the graded A_G–module

$$H_G(X, F) := H(p_* F).$$

13.2. Definition. Assume that X is nice (1.4) and let $F \in D_G^b(X)$. The G–equivariant cohomology $H_{G,c}(X, F)$ with **compact supports** of X with coefficients in F is by definition the graded A_G–module

$$H_{G,c}(X, F) := H(p_! F).$$

13.3. Certainly the object $p_* F \in D_G^+(pt)$ (or $p_! F$) carries more information than the A_G–module $H_G(X, F)$ (or $H_{G,c}(X, F)$), and we usually prefer to work with the triangulated category $D_G^+(pt)$ rather than with the abelian one Mod_{A_G}. In particular, if X is a pseudomanifold and we work with the constructible category $D_{G,c}^b(X)$, then we interpret the formula

$$D \cdot p_! \simeq p_* \cdot D$$

as the **equivariant Poincare duality**. Notice that in case of a connected Lie group G this formula relates the Verdier duality in $D_{G,c}^b(X)$ with the "coherent" duality in $D_{A_G}^f \simeq D_{D,c}^b(pt)$ (see 12.7.2(ii)).

For the rest of this section 13 we put $R = \mathbb{R}$.

13.4. Example. Let G be a complex linear algebraic group acting algebraically on a complex algebraic variety X. Consider the G–equivariant intersection cohomology sheaf $IC_G(X)$. We denote by

$$IH_G(X) := H_G(X, IC_G(X))$$

$$IH_{G,c}(X) := H_{G,c}(X, IC_G(X))$$

the **equivariant intersection cohomology** (resp. with compact supports) of X.

Assume that X is proper. Then by the decomposition theorem (5.3) the direct image $p_*IC_G(X) = p_!IC_G(X)$ is a direct sum of (shifted) local systems on BG. If furthermore the group G is connected then each local system is constant, and we conclude that $IH_G(X)$ is a *free* A_G-module with the graded basis $IH(X)$, i.e.

$$IH_G(X) = A_G \otimes IH(X).$$

13.5. Let $\phi : H \to G$ be a homomorphism of Lie groups. It induces a map of classifying spaces $\bar{\phi} : BH \to BG$ and hence a map on cohomology $A_G \to A_H$. For an A_H-module M we denote by $_{A_G}M$ the corresponding A_G-module obtained via the restriction of scalars. It is clear that for $S \in D^+(BH)$ we have

$$(*) \qquad\qquad H(\bar{\varphi}_*S) = {}_{A_G}H(S).$$

Let $f : X \to Y$ be a ϕ-map and $F \in D_H^+(X)$. Let $Q_{f_*}F \in D_G^+(Y)$ be its direct image. The following formula immediately follows from $(*)$ and 6.12.2

$$_{A_G}H_H(X, F) = H_G(Y, Q_{f_*}F).$$

13.6. Example. Let $f : X \to Y$ be a principal G–bundle. Then we know that $Q_*C_{X,G} = C_Y$. Hence by the above formula we have an isomorphism of graded groups

$$H_G(X) = H(Y).$$

Of course, $H_G(X)$ has more structure, namely the action of generators of A_G (the Chern classes of the G–bundle $X \to Y$).

13.7. Example. In the previous example assume that Y is a compact manifold and $G = S^1$. Assume moreover that the manifold X is orientable. We want to make the equivariant Poincare duality explicit in this case. Consider the map $p : X \to pt$. We have $A_{S^1} \simeq \mathbb{R}[x]$ and identify $D^b_{S^1, c}(pt) = D^f_{A_{S^1}}$ (12.7.2). Since X is compact we have $p_*C_{X,S^1} = p_!C_{X,S^1}$. Also $DC_{X,S^1} = C_{X,S^1}[d_X]$, $d_X = \dim X$, since X is orientable. Hence the Poincare duality formula

$$D \cdot p_! C_{X,S^1} = p_* \cdot DC_{X,S^1}$$

becomes

$$D \cdot p_* C_{X,S^1} = p_* C_{X,S^1}[d_X].$$

Put $p_*C_{X,S^1} = M \in D^f_{A_{S^1}}$. Since $H(M)$ has cohomological dimension ≤ 1 as an A_{S^1} – module, it follows from 11.3.3 that $M = H(M)$, i.e., the DG–module M has the zero differential. Note that in fact M is a torsion module, since $M = H(Y, \mathbb{R})$. Hence by 11.3.1(i) we have

$$DM = \text{Ext}^1_{A_{S^1}}(M, A_{S^1})[-1]$$

So the Poincare duality is a canonical isomorphism for the A_{S^1} – module $H_{S^1}(X)$

$$H_{S^1}(X) = \text{Ext}^1_{A_{S^1}}(H_{S^1}(X), A_{S^1})[-d_X - 1]$$

13.8. Remark. Replace in the previous example S^1 by an arbitrary connected compact Lie group K of rank r. If we knew that the DG–module $p_* C_{X,K} \in D^f_{A_K}$ had zero differential (which is probably true), then we would obtain the similar formula

$$H_K(X) = \text{Ext}^r_{A_K}(H_K(X), A_K)[-d_X - r]$$

using the same argument.

13.9. Example. Suppose we are in the situation of Theorem 9.1. Namely, let $0 \to K \to H \xrightarrow{\varphi} G \to 0$ be an exact sequence of complex linear reductive algebraic groups. Let $f : X \to Y$ be an algebraic morphism which is a ϕ-map. Assume that the following conditions hold.

(a) The group K acts on X with only finite stabilizers.

(b) The morphism f is affine and is the geometric quotient map by the action of K (all K-orbits on X are closed).

Then we know that $Q_* IC_H(X) = IH_G(Y)[d_K]$, where $d_K = \dim_{\mathbb{C}} K = d_X - d_Y$. Hence as in 13.5 above we obtain an isomorphism of A_G-modules

$$_{A_G} IH_H(X) = IH_G(Y)[d_X - d_Y].$$

13.10. Borel's interpretation of A_G.

Let G be a compact Lie group and T be a maximal torus in G. Let t be the Lie algebra of T and W be the Weyl group $W = N(T)/T$. The group W acts on t and hence also on the ring of polynomial functions $S(t^*)$ on t. By the classical result of A. Borel ([Bo3]) we have a canonical isomorphism of graded algebras

$$A_G = H(BG) \simeq S(t^*)^W,$$

where linear functions in $S(t^*)$ are assigned degree 2.

If G is connected then $S(t^*)^W$ is a polynomial ring and $S(t^*)$ is a *free* $S(t^*)^W$– module, since the group W is generated by reflections. Let $U \subset G$ be a closed subgroup with a maximal torus $T' \subset U$ and the Weyl group W'. The diagram of inclusions of groups

$$\begin{array}{ccc} T' & \to & T \\ \downarrow & & \downarrow \\ U & \to & G \end{array}$$

induces the diagram of classifying spaces

$$
\begin{array}{ccc}
BT' & \to & BT \\
\downarrow & & \downarrow \\
BU & \to & BG
\end{array}
$$

and the corresponding diagram of cohomology rings

$$
\begin{array}{ccc}
A_{T'} & \leftarrow & A_T \\
\uparrow & & \uparrow \\
A_U & \leftarrow & A_G.
\end{array}
$$

Borel showed that this diagram coincides with the natural diagram

$$
\begin{array}{ccc}
S(t'^*) & \leftarrow & S(t^*) \\
\uparrow & & \uparrow \\
S(t'^*)^{W'} & \leftarrow & S(t^*)^W
\end{array}
$$

under the above identification (the horizontal arrows in the last diagram are restrictions of functions).

Let G be a Lie group with finitely many components and $K \subset G$ be a maximal compact subgroup. Then topolically

$$
G \simeq K \times \mathbb{R}^d.
$$

Hence $A_G = A_K$ and the above picture can be applied to A_G.

13.11. Equivariant cohomology of induced spaces.

13.11.1. Let G be a group and $\phi : H \hookrightarrow G$ be an embedding of a closed subgroup. Let X be an H–space and $Y = G \times_H X$ be the induced G–space. The inclusion $f : X \hookrightarrow Y$ is a ϕ–map. Let $F \in D_H^+(X)$ and $Q_{f*}F \in D_G^+(Y)$. Then by 13.5 we have

$$
_{A_G}H_H(F) = H_G(Q_{f*}F).
$$

We want to derive a similar relation for the cohomology with compact supports. So let us assume in addition that X is a constructible space (1.10) and that H, G are Lie groups with finitely many components.

Consider the commutative functorial diagram

$$
\begin{array}{ccc}
D_{H,c}^b(X) & \xrightarrow{Q_{f*}} & D_{G,c}^b(Y) \\
p_* \downarrow & & \downarrow p_* \\
D_{H,c}^b(pt) & \xrightarrow{Q_*} & D_{G,c}^b(pt).
\end{array}
$$

In order to find a relation between $H_{H,c}(F)$ and $H_{G,c}(Q_{f*}F)$ we need to know how functors Q_{f*}, Q_* behave with respect to duality.

We denote by d_M the dimension of a manifold M.

13.11.2. Recall (7.6.3) that if the group H is connected then there exists a canonical isomorphism of functors from $D^b_{H,c}(X)$ to $D^b_{G,c}(Y)$

$$Q_{f*} \cdot D = (D \cdot Q_{f*})[d_H - d_G].$$

Denote by $K(H) \subset H$ and $K(G) \subset G$ the maximal compact subgroups of H, G.

13.11.3. Proposition. *Assume that $K(H)$ is connected. Then there exists a canonical isomorphism of functors*

$$Q_* \cdot D = (D \cdot Q_*)[d_{K(H)} - d_{K(G)}].$$

Proof. We may (and will) assume that $K(H) = H, K(G) = G$.

Let E be an ∞-acyclic free G-space, hence also a free H-space. Then $G\backslash E = BG$, $H\backslash E = BH$ and we have the natural fibration $\pi : BH \to BG$ with the fiber G/H. Recall that categories $D^b_{H,c}(pt)$ and $D^b_{G,c}(pt)$ are naturally identified as certain full subcategories in $D^b(BH)$ and $D^b(BG)$ respectively. Under this identification the functor Q_* is π_*.

Consider the G-space $Z = G/H$ and its ∞-acyclic free resolution $P = E \times Z \to Z$. let $\overline{P} = G\backslash P$. Then $D^b_{G,c}(Z)$ is identified as a full subcategory in $D^b(\overline{P})$. Notice that $\overline{P} = H\backslash E = BH$ and the categories $D^b_{G,c}(Z)$ and $D^b_{H,c}(pt)$ are identified as the same full subcategory in $D^b(\overline{P}) = D^b(BH)$. Indeed, both categories consist of bounded complexes $S \in D^b(\overline{P})$ with constant cohomology sheaves of finite rank. Moreover, the direct image functor $\pi_* : D^b_{H,c}(pt) \to D^b_{G,c}(pt)$ is then identified with the direct image $p_* : D^b_{G,c}(Z) \to D^b_{G,c}(pt)$ for the G-map $p : Z \to pt$.

The space Z is compact, hence $p_* = p_!$ and so by 13.3

$$D \cdot p_* = p_* \cdot D,$$

where the duality D on the right takes place in $D^b_{G,c}(Z)$. So it remains to show that the equivalence of categories $D^b_{G,c}(Z) \to D^b_{H,c}(pt)$ commutes with the duality up to the shift by $d_Z = d_G - d_H$. Since the group H is connected this follows from 7.6.3. This proves the proposition.

We can now state our main result.

13.11.4. Theorem. *Assume that in the setup of 13.11.1 the group H is connected. There exists a natural isomorphism of functors*

$$Q_* \cdot p_! = p_! \cdot Q_{f*}[d_{K(H)} - d_{K(G)} + d_G - d_H]$$

from $D^b_{H,c}(X)$ to $D^b_{G,c}(pt)$.

In particular for $F \in D^b_{H,c}(X)$ we have a natural isomorphism of A_G-modules

$$A_G H_{H,c}(F) = H_{G,c}(Q_{f*}F)[d_{K(H)} - d_{K(G)} + d_G - d_H]$$

Proof. Recall that $p_! = D \cdot p_* \cdot D$. So it remains to apply 13.11.2 and 13.11.3.

13.12. Relation with nonequivariant cohomology.

13.12.1. Let G be a Lie group and X be a G–space. Let $\phi : H \hookrightarrow G$ be an embedding of a closed subgroup. We want to compare the G–equivariant cohomology of X with the H–equivariant one.

Consider the commutative diagram

$$
\begin{array}{ccc}
X & \stackrel{f=id}{\to} & X \\
p \downarrow & & \downarrow p \\
pt & \stackrel{\pi}{\to} & pt
\end{array}
$$

where horizontal arrows are ϕ–maps. By theorem 7.3 we have a natural isomorphism of functors

$$p_* \cdot Q_f^* = Q_\pi^* \cdot p_*$$

from $D_G^+(X)$ to $D_H^+(pt)$.

13.12.2. Corollary. *In the setup of 13.12.1 assume that the groups G, H are connected and identify $D_G^+(pt) = D_{\mathcal{A}_G}^+$, $D_H^+(pt) = D_{\mathcal{A}_H}^+$ (12.7.2(i),(iii)). Then for $F \in D_G^+(X)$ we have*

$$p_* \cdot Res_{H,G}F = \mathcal{A}_H \overset{L}{\otimes}_{\mathcal{A}_G} (p_* F).$$

In particular, if $H = \{e\}$, then the nonequivariant cohomology $H(X, For(F))$ is computed from $p_* F \in D_{\mathcal{A}_G}^+$ by the formula

$$H(X, For(F)) = \mathbb{R} \overset{L}{\otimes}_{\mathcal{A}_G} (p_* F).$$

13.12.3. Remark. Let $F \in D_G^+(X)$. If we only know the equivariant cohomology $H_G(X, F)$ then we cannot in general recover the nonequivariant one $H(X, For(F))$. However, we have more information if we work with the DG–module $p_* F \in D_{\mathcal{A}_G}^+$. If, for example $p_* F \in D_{\mathcal{A}_G}^f$, then $H(X, For(F))$ is a graded basis of a minimal $\mathcal{K}$–projective $\mathcal{A}_G$–module P quasiisomorphic to $p_* F$ (11.4.6).

14. Fundamental example.

We analyze the stalk of the equivariant intersection cohomology sheaf at a point fixed by a 1-parameter subgroup.

14.1. Let $0 \to \mathbb{C}^* \to H \xrightarrow{\phi} G \to 0$ be an exact sequence of complex connected reductive groups. Let X be an affine complex variety with an algebraic action of the group H. Assume that X has an H-fixed point q which is the unique $\mathbb{C}^*$-fixed point. Assume that q is the attraction point under the $\mathbb{C}^*$-action on X, that is, the ring of functions on X is nonnegatively graded by characters of $\mathbb{C}^*$.

Let $\{q\} \xhookrightarrow{i} X \xhookleftarrow{j} X_0 := X - \{q\}$ denote the corresponding closed and open embeddings. Put $F := IC_H(X)$ and $F_q^! := i^! F \in D_H^b(\{q\})$, $F_0 := j^* F \in D_H^b(X_0)$. Consider the exact triangle in $D_{H,c}^b(X)$

$$(1) \qquad\qquad i_* F_q^! \to F \to j_* F_0$$

and its direct image in $D_{H,c}^b(pt)$ under the map $p : X \to pt$

$$(2) \qquad\qquad p_* F_q^! \to p_* F \to p_* F_0.$$

We identify the categories $D_{H,c}^b(pt) = D_{\mathcal{A}_H}^f$ (12.7.2(ii)) and denote the DG–module $p_* F_0$ by M. Consider the canonical exact triangle in $D_{\mathcal{A}_H}^f$

$$(3) \qquad\qquad (\tau_{\geq 0} M)[-1] \to \tau_{<0} M \to M.$$

14.2. Theorem.
(i) The triangles (2), (3) above are isomorphic.

(ii) The objects $\tau_{<0} M$, $\tau_{\geq 0} M \in D_{\mathcal{A}_H}^f$ are free $\mathcal{A}_H$-modules with zero differential and a basis given by $IH(X)$ and the costalk $i^! IC(X)$ respectively.

The proof of this theorem uses the decomposition theorem and the hard Lefschetz theorem. Let us first of all deduce an important corollary.

14.3. Corollary. *(i) The costalk $F_q^!$ is a **direct sum** of (shifted) constant equivariant sheaves $C_{q,H}$ at q:*

$$F_q^! = C_{q,H} \otimes i^! IC(X)$$

(i') Similarly for the stalk $F_q = i^* F$:

$$F_q = C_{q,H} \otimes i^* IC(X)$$

(ii) The equivariant intersection cohomology $IH(X)$ is a free $\mathcal{A}_H$-module with a basis $IH(X)$, i.e.

$$IH_H(X) = \mathcal{A}_H \otimes IH(X)$$

(ii') *Similarly for $IH_{H,c}(X)$:*

$$IH_{H,c}(X) = A_H \otimes IH_c(X)$$

Proof. (i) and (ii) follow immediately from the theorem; (i') and (ii') follow from (i) and (ii) by duality, since all basic functors commute with the forgetful functor.

14.4. Proof of theorem 14.2.

By our assumptions the action of the subgroup $\mathbb{C}^* \subset H$ defines on X the structure of an affine quasihomogeneous cone over the projective variety $\overline{X} = \mathbb{C}^* \backslash X_0$. Note that the group G acts naturally on $\overline{X}$ and the projection $f : X_0 \to \overline{X}$ is a ϕ–map.

(i) Let $Q_0, Q_{-1} \in D^f_{A_H}$ be minimal $\mathcal{K}$–projective DG–modules quasiisomorphic to $p_* F$ and $p_* F^!_q$ respectively, so that the triangle (2) is isomorphic to a triangle

$$(2') \qquad\qquad Q_{-1} \overset{\varepsilon}{\to} Q_0 \to M$$

By remark 13.12.3 the free A_H–modules Q_0, Q_{-1} have bases $IH(X)$ and $i^! IC(X)$ respectively. It is known that $IH(X)[-1]$ and $i^! IC(X)[1]$ are isomorphic to the primitive and the coprimitive parts of $IH(\overline{X})$ with respect to the Lefschetz operator on $\overline{X}$. Hence is particular Q_0 is generated in degrees < 0 and Q_{-1} is generated in degrees > 0. This implies that in the $\mathcal{K}$–projective module $Q = \operatorname{cone}(\varepsilon) = Q_0 \oplus Q_{-1}[1]$ we have

$$d_Q Q \subset mQ,$$

where $m \subset A_H$ is the maximal ideal. Therefore Q is the *minimal* $\mathcal{K}$–projective quasisomorphic to M (11.4.7). Moreover,

$$Q_0 = \tau_{<0} M$$

$$Q_{-1} = (\tau_{\geq 0} M)[-1].$$

Hence triangles (2') and (3) are isomorphic. This proves (i).

(ii) The homomorphism $\phi : H \to G$ induces an embedding $A_G \hookrightarrow A_H$. We have (non canonically)

$$A_H \simeq A_G[\lambda]$$

where λ has degree 2.

Consider the ϕ–map $f : X_0 \to \overline{X}$. Since $\mathbb{C}^*$ acts with only finite stabilizers, $Q_{f_*} F_0 = IC_G(\overline{X})[1]$ (9.1(iv)). Therefore

$$(*) \qquad\qquad _{A_G} M = IH_G(\overline{X})[1]$$

(13.5). The variety $\overline{X}$ is projective, so $IH_G(\overline{X})$ is a free A_G-module and

$$IH_G(\overline{X}) = A_G \otimes_R IH(\overline{X})$$

(13.4). Therefore we obtain an isomorphism of $\mathbb{R}$-vector spaces

(4) $$\mathbb{R} \otimes_{A_G} M \simeq IH(\overline{X})[1].$$

Notice that the left hand side in (4) is naturally an $\mathbb{R}[\lambda]$-module.

14.5. Lemma. *Under the identification (4) the action of λ on $IH(\overline{X})$ coincides (up to a scalar) with the Lefschetz operator for the projective variety $\overline{X}$.*

Let us postpone the proof of the lemma and finish the proof of the theorem.

Since the A_G-module $_{A_G}H(M) = IH_G(\overline{X})[1]$ is free, ,the A_H-module $H(M)$ has cohomological dimension ≤ 1. So by proposition 11.3.3 the DG-module M has zero differential $M = H(M)$.

Let Pr and CPr be the primitive and the coprimitive parts of the cohomology $IH(\overline{X})$ with respect to the Lefschetz operator λ. By considering the Hilbert polynomial of the A_H-module M we find that it has a minimal projective resolution of the form

$$0 \to P_{-1} \xrightarrow{\delta} P_0 \to M \to 0,$$

where $P_0 = A_H \otimes_R Pr[1]$, $P_{-1} = A_H \otimes_R CPr[-1]$. Hence $\mathrm{Cone}(\delta) = P_0 \oplus P_{-1}[1]$ is a minimal $\mathcal{K}$-projective quasiisomorphic to M. Moreover

$$P_0 = \tau_{<0} M$$

$$P_{-1} = (\tau_{\geq 0} M)[-1]$$

(use Hard Lefschetz for $\overline{X}$). Hence $P_0 = Q_0, P_{-1} = Q_{-1}$ and hence Q_0, Q_{-1} have zero differential which proves part (ii) in the theorem.

Proof of Lemma 14.5.

Consider the commutative diagram of group homomorphisms

$$\begin{array}{ccc} \mathbb{C}^* & \to & H \\ \downarrow & & \downarrow \\ \{e\} & \to & G. \end{array}$$

Since $\phi : H \to G$ is surjective the assumption of theorem 7.3 is satisfied. Consider the diagram

$$\begin{array}{ccc} X_0 & \xrightarrow{f} & \overline{X} \\ p\downarrow & & \downarrow p \\ pt & \xrightarrow{\pi} & pt \end{array}$$

where the horizontal arrows are the ϕ-maps. Then by theorem 7.3 and proposition 7.2 the functors of the direct image $Q_{f*}, Q_{\pi*}, p_*$ commute with the restriction functors $Res_{\{e\}, G}$,
$Res_{\mathbb{C}^*, H}$. We have to prove something about the object

$$Res_{\{e\}, G} \cdot Q_{\pi*} \cdot p_* F_0 = \mathbb{R} \otimes_{A_G} M \in D^+_{\{e\}}(pt).$$

But by above mentioned results it is equal to

$$Q_{\pi*} \cdot p_* \cdot Res_{\mathbb{C}^*, H} F_0.$$

So we may (and will) assume that $G = \{e\}, H = \mathbb{C}^*$.

Let E be an ∞-acyclic free $\mathbb{C}^*$-space and $\mathbb{C}^* \backslash E = B\mathbb{C}^*$ be the classifying space. Put $X_{0\mathbb{C}^*} = \mathbb{C}^* \backslash (X_0 \times E)$. We have natural projections

(1)
$$\overline{X} \xleftarrow{\overline{f}} X_{0\mathbb{C}^*} \xrightarrow{p} B\mathbb{C}^*.$$

If we embed $D^+_{\mathbb{C}^*}(X_0) \subset D^+(X_{0\mathbb{C}^*})$ then the direct image $Q_{f*} : D^+_{\mathbb{C}^*}(X_0) \to D^+(\overline{X})$ becomes $\overline{f}_* : D^+(X_{0\mathbb{C}^*}) \to D^+(\overline{X})$. We know that $\overline{f}_* C_{X_{0\mathbb{C}^*}} = C_{\overline{X}}$ (9.1(ii)), hence

(2)
$$H(\overline{X}) = H(X_{0\mathbb{C}^*}).$$

But $H(X_{0\mathbb{C}^*})$ is a module over $H(B\mathbb{C}^*) = \mathbb{R}[\lambda]$ via the projection p in (1); and the image of λ on $H(\overline{X})$ via the identification (2) above is the first Chern class of the (almost) principal $\mathbb{C}^*$-bundle $f : X_0 \to \overline{X}$. This proves the lemma.

14.6. Corollary. *In the previous setup the natural map of A_H-modules*

$$IH_H(X) \to IH_H(X_0)$$

induces an isomorphism modulo the maximal ideal

$$IH(X) = \mathbb{R} \otimes_{A_H} IH_H(X) \xrightarrow{\sim} \mathbb{R} \otimes_{A_H} IH_H(X_0)$$

Proof. Indeed, in the proof of the above theorem 14.2 we saw that this map is the minimal projective cover

$$P_0 \to M.$$

Hence the assertion follows.

14.7. Consider the dual picture.

Put $F_q := i^* F$ and consider the exact triangle

(1')
$$F[-1] \to i_* F_q[-1] \to j_! F_0$$

which is the dual to the triangle (1) in 14.1. Consider its direct image with compact supports in $D^b_{H,c}(pt)$

$$(2') \qquad\qquad p_! F[-1] \to p_! F_q[-1] \to p_! F_0.$$

Since $DF = F$ we find that the triangle $(2')$ is dual to the triangle (2) in 14.1.

By the above theorem 14.2 the triangle (2) in 14.1 is of the form

$$(3) \qquad\qquad (\tau_{\geq 0} M)[-1] \to \tau_{<0} M \to M$$

where all modules have zero differential. Moreover, the A_H-modules $\tau_{\geq 0} M, \tau_{<0} M$ are free and the diagram (3) is the minimal projective resolution of the A_H-module M. Hence the dual traingle is

$$(3') \qquad\qquad \tau_{>0}(DM)[-1] \to \tau_{\leq 0}(DM) \to DM,$$

which is also a minimal projective resolution of the A_H-module $DM = \mathrm{Ext}^1_{A_H}(M, A_H)$ (11.3.1(i)). Identifying terms in isomorphic triangles $(2')$ and $(3')$ we find the following

14.8. Corollary. *The natural map of A_H-modules*

$$(IC_{X,H})_q[-1] \to IH_{H,c}(X_0)$$

induces an isomorphism modulo the maximal ideal

$$IC_{X,q}[-1] = \mathbb{R} \otimes_{A_H} (IC_{X,H})_q[-1] \overset{\sim}{\to} \mathbb{R} \otimes_{A_H} IH_{H,c}(X_0).$$

Part III. Equivariant cohomology of toric varieties.

15. Toric varieties.

15.0. In this part we present some applications of the theory developed in parts I and II. Namely, we work out the "simpliest" case of toric varieties. It turns out that there exists a natural complex which is a resolution of the equivariant intersection cohomology (with compact supports) of a toric variety. As a byproduct we obtain some applications to combinatorics (see theorem 15.7 below).

We start by recalling the notion of a toric variety. Then we define some combinatorial structure (the minimal complex) on the fan corresponding to a toric variety. After that we state our main result about this structure and its relation with the equivariant intersection cohomology. The rest of the text is devoted to the proof of the main theorem 15.7.

15.1. Review of toric varieties.

We recall some basic notions and results in the theory of toric varieties. For the proofs the reader is referred to [Dan] or [KKMS-D].

Let $T = (\mathbb{C}^*)^n$ be a complex torus of dimension n. Let $\Lambda := \operatorname{Hom}(\mathbb{C}^*, T) \simeq \mathbb{Z}^n$ be the group of 1–parameter subgroups in T. Denote by $N = \Lambda \otimes_{\mathbb{Z}} \mathbb{R}$ the corresponding real vector space of dimension n. Note that N can be identified with the Lie algebra of the compact torus $(S^1)^n \subset (\mathbb{C}^*)^n$.

Let $Y = Y^n$ be an algebraic variety. We say that Y is a **toric variety** (or a T–toric variety) if T acts on Y and Y contains a dense orbit isomorphic to T (this implies that the number of T–orbits in Y is finite).

A normal toric variety Y is described combinatorially by a **fan** $\Phi_Y = \Phi$ in N. Recall that a fan Φ is a collection $\Phi = \{\sigma\}$ of finitely many **rational** (with respect to the lattice $\Lambda \subset N$) **convex polyhedral cones** σ which intersect along common faces, such that if $\sigma \in \Phi$ and τ is a face of σ then also $\tau \in \Phi$.

In that description Y is **affine** if Φ consists of a unique cone σ together with its faces and X is **complete** if $\cup_{\sigma \in \Phi} \sigma = N$.

The orbits $\mathcal{O}$ of T in Y are in $1-1$ correspondence with the cones $\sigma \in \Phi$. A cone $\sigma \in \Phi$ corresponding to an orbit $\mathcal{O}$ will be denoted by $\sigma_{\mathcal{O}}$, and vice versa, we denote by $\mathcal{O}_\sigma$ the orbit corresponding to a cone $\sigma \in \Phi$. We have $\dim_{\mathbb{C}} \mathcal{O} = n - \dim_{\mathbb{R}}(\sigma_{\mathcal{O}})$. More precisely, the subspace of N spanned by $\sigma_{\mathcal{O}}$ is the (real part of) Lie algebra of the stabilizer of $\mathcal{O}$. We have $\mathcal{O}' \subset \overline{\mathcal{O}}$ if and only if $\sigma_{\mathcal{O}} \subset \sigma_{\mathcal{O}'}$.

15.2. Fix a torus $T = (\mathbb{C}^*)^n$. Let $A = \mathbb{R}[x_1, \ldots, x_n]$ be the ring of polynomial functions on N. We consider A as a graded ring, where $\deg(x_i) = 2$. Let $m \subset A$ be the maximal ideal. For a cone $\sigma \subset N$ denote by A_σ the (graded) ring of polynomial functions on σ. The restriction of functions defines the ring homomorphism $A \to A_\sigma$. Fix a fan $\Phi = \{\sigma\}$ in N (not necessarily rational).

Consider a complex

$$\mathcal{K} : 0 \to \mathcal{K}^{-n} \overset{\partial^{-n}}{\to} \mathcal{K}^{-n+1} \overset{\partial^{-n+1}}{\to} \cdots \overset{\partial^{-1}}{\to} \mathcal{K}^0 \to 0$$

of graded A–modules, where $\mathcal{K}^{-i}$ is a direct sum of terms $\mathcal{K}_\sigma$ for $\sigma \in \Phi$, $\dim \sigma = i$:

$$\mathcal{K}^{-i} = \oplus_{\dim \sigma = i} \mathcal{K}_\sigma.$$

Assume that each $\mathcal{K}_\sigma$ is a free A_σ–module, i.e., $\mathcal{K}_\sigma$ is a direct sum of (shifted) topics of A_σ.

We assume that the differential ∂ maps $\mathcal{K}_\sigma$ to $\oplus_{\tau \subset \sigma} \mathcal{K}_\tau$. Let $\sigma \in \Phi$ be a dimension i. Denote by

$$\mathcal{K}(\sigma) : \mathcal{K}_\sigma \overset{\partial_\sigma^{-i}}{\to} \underset{\substack{\tau \subset \sigma \\ \dim \tau = i-1}}{\oplus} \mathcal{K}_\tau \overset{\partial_\sigma^{-i+1}}{\to} \cdots \to \mathcal{K}^0 \to 0$$

the "restriction" of the complex $\mathcal{K}$ to the cone σ with its faces.

15.3. Definition. A complex $\mathcal{K}$ as above is called **minimal** if it satisfies the following conditions

(a) $\mathcal{K}^0 = \mathbb{R}[n]$, i.e. it is the A–module $\mathbb{R} = A/m$ placed in degree $-n$.

(b) Let $I_\sigma = \ker \partial_\sigma^{-i+1}$ in the complex $\mathcal{K}(\sigma)$. Then the differential ∂_σ^{-i} induces an isomorphism of $\mathbb{R}$–vector spaces

$$\partial_\sigma^{-i} : \mathcal{K}_\sigma / m\mathcal{K}_\sigma \overset{\sim}{\to} I_\sigma / m I_\sigma.$$

One can construct a minimal complex by induction on the dimension of σ.

15.4. Lemma. *Let $\mathcal{K}', \mathcal{K}''$ be two minimal complexes. Then they are isomorphic (noncanonically).*

Proof. The proof is essentially the same as that of the uniqueness of a minimal projective resolution. Namely, one constructs an isomorphism between $\mathcal{K}'$ and $\mathcal{K}''$ step by step by induction on the dimension of σ.

15.5. Remark. A minimal complex $\mathcal{K}$ is by definition "locally" exact (except in degree $-n$), i.e., $im\, \partial_\sigma^{-i} = \ker \partial_\sigma^{-i+1}$ for all σ. Hence the exactness of $\mathcal{K}$ at $\mathcal{K}^{-i}$ is equivalent to the kernel $\ker \partial^{-i}$ being the sum of local kernels $\ker \partial_\sigma^{-i}$, $\dim \sigma = i$.

15.6. Note that the ring A is canonically identified with the cohomology ring $A_T = H(BT, \mathbb{R})$ (13.10). Hence in case of a rational fan Φ one may hope that the minimal complex has a meaning in terms of the T–equivariant cohomology of the corresponding toric variety X. This is so indeed.

15.7. Theorem. *Let Φ be a rational fan in N, corresponding to a normal toric variety X. Assume that Φ is complete or consists of a single cone of dimension n together with its faces. Then the following hold.*

(i) The minimal complex is **exact** *except in degree $-n$.*

(ii) There is an isomorphism of A-modules

$$Ker\, \partial^{-n} \simeq IH_{T,c}(X)$$

(iii) The free A_σ-module $\mathcal{K}_\sigma$ has a graded basis isomorphic to the stalk $IC(X)|_{\mathcal{O}_\sigma}$ of the intersection cohomology sheaf $IC(X)$ on the corresponding orbit $\mathcal{O}_\sigma \subset X$.

15.8. Remarks. 1. The assumptions on the fan Φ in the theorem mean that the corresponding variety X is complete or affine with a (unique) T-fixed point. In these cases we know that $IH_{T,c}(X)$ is a **free** A-module with a basis $IH_c(X)$ (see 13.4 and 14.3(ii')). Hence by (ii) in the theorem we get

$$Ker\, \partial^{-n} \simeq A \otimes IH_c(X).$$

2. Assume that the fan Φ is simplicial. Then the variety X has only quotient singularities, i.e., $IC(X) = C_X$. Then from (iii) we get $\mathcal{K}_\sigma \simeq A_\sigma$. Hence a minimal complex is isomorphic to the complex of functions, i.e. $\mathcal{K}_\sigma = A_\sigma$ and ∂ is the restriction of functions with $\pm$ sign depending on some chosen orientation of the cones σ. Can the reader check directly that this complex of functions is exact except at $\mathcal{K}^{-n}$?

3. We will prove the theorem in two steps. First of all we will construct a certain **canonical** "geometric" complex $\mathcal{L}$ of A-module coming from the variety X and prove that it is a resolution of $IH_{T,c}(X)$. Then we will prove that $\mathcal{L}$ is a minimal complex. We know no other way of proving that a minimal complex is exact (away from $\mathcal{K}^{-n}$) except by interpreting it geometrically as the complex $\mathcal{L}$.

4. The complex $\mathcal{L}$ mentioned above provides a **natural** resolution of $IH_{T,c}(X)$. Assume, for example, that a group Γ acts on the space N by automorphisms of the lattice Λ and preserves the fan Φ. Then Γ also acts on the toric variety X and hence on its cohomology $IH_c(X)$. It is sometimes easy to compute the (graded) character of Γ on each term $\mathcal{L}^{-i}$ of the complex $\mathcal{L}$ (for example, if the fan Φ is simplicial). This provides the charater of Γ on the equivariant intersection cohomology $IH_{T,c}(X)$. But $IH_{T,c}(X)$ is a *free* A-module with the basis $IH_c(X)$. Hence from the character of Γ on $IH_{T,c}(X)$ one can find the character of Γ on $IH_c(X)$.

Here is an example. Let T be a maximal torus in an algebraic group G. We have the (simplicial) fan Φ of the Weyl chambers in N. The Weyl group W acts on the fan Φ and hence on the corresponding toric variety X. The above method allows one to compute the character of W on the cohomology $H(X)$ (see [DL]).

15.9. Conjecture. The statement (i) in the theorem still holds if we drop the assumption on the rationallity of the fan Φ.

15.10. Proof of Theorem 15.7.

Consider the following filtration of X by closed subsets

$$\emptyset = X^{-1} \subset X^0 \subset X^1 \subset \ldots \subset X^n = X,$$

where $X^k := \coprod_{dim\mathcal{O} \le k} \mathcal{O}$, $k = -1, 0, 1, \ldots, n$.

Put $U^{k+1} := X - X^k$, so that $U^{n+1} = \emptyset$, $U^0 = X$ and $U^n = T$ is the dense orbit in X. Let $Z^k := X^k - X^{k-1} = \coprod_{dim\mathcal{O}=k} \mathcal{O}$.

Denote by

$$j_k : U^k \hookrightarrow X,$$

$$i_k : Z^k \hookrightarrow X$$

the open and the locally closed embeddings.

Put $F = IC_T(X)$ – the T–equivariant intersection cohomology sheaf on X, and

$$F_k := j_{k!} j_k^* F.$$

Consider the collection of exact triangles in $D^b_{T,c}(X)$:

(1) $$F_k[k] \to i_{k!} i_k^* F[k] \to F_{k+1}[k+1]$$

for $k = 0, \ldots, n$.

Apply the functor $H_{T,c}$ – equivariant cohomology with compact supports – to triangles (1). This produces complexes of A_T–modules

(2) $$0 \to H_{T,c}(F_k)[k] \xrightarrow{a_k} H_{T,c}(i_k^* F)[k] \xrightarrow{b_k} H_{T,c}(F_{k+1})[k+1] \to 0$$

Consider the induced complex

(3) $$0 \to H_{T,c}(i_0^* F) \xrightarrow{d^0} H_{T,c}(i_1^* F)[1] \xrightarrow{d^1} \cdots \xrightarrow{d^{n-1}} H_{T,c}(i_n^* F)[n] \to 0$$

where $d^i = a_{i+1} \cdot b_i$.

15.11. Theorem. *(a) The complex (3) above is exact except at $H_{T,c}(i_0^* F)$.*

(b) $Ker d^0 = IH_{T,c}(X)$.

This theorem is equivalent to the following

15.12. Claim. *Complexes (2) are short exact sequences.*

It is easier to prove a more precise statement. We need the following

15.13. Definition. An A–module $P \neq 0$ is called (cohomologically) pure of codimension k if

$$\mathrm{Ext}_A^i(P, A) = 0 \text{ for } i \neq k.$$

The theorem follows from the following

15.14. Proposition. *The sequences (2) are short exact and the A-module $H_{T,c}(F_k)$ is pure of codimension k.*

Proof of proposition 15.14.

15.15. Lemma. *Let $\mathcal{O} \subset X$ be an orbit. Then the restriction $F|_{\mathcal{O}}$ is a direct sum of (shifted) constant equivariant sheaves $C_{\mathcal{O},T}$.*

Proof of lemma. Let $U \subset X$ be the star of the orbit $\mathcal{O}$, that is, U is an open set consisting of orbits $\mathcal{O}'$ such that $\mathcal{O} \subset \overline{\mathcal{O}^i}$. Let $T_{\mathcal{O}} = Stab(\mathcal{O}) \subset T$ be the stabilizer of the orbit $\mathcal{O}$. Since X in normal, one can show that $T_{\mathcal{O}}$ is connected hence a subtorus in T. Let $X_{\mathcal{O}} = \overline{T}_{\mathcal{O}} \subset U$ be the closure in U of the subtorus $T_{\mathcal{O}} \subset T \subset U$. Then the T–space U is the induced one from the $T_{\mathcal{O}}$–space $X_{\mathcal{O}}$

$$U = T \times_{T_{\mathcal{O}}} X_{\mathcal{O}}.$$

The space $X_{\mathcal{O}}$ is an affine $T_{\mathcal{O}}$-toric variety with a fixed point $q = X_{\mathcal{O}} \cap \mathcal{O}$. Hence it satisfies assumptions of theorem 14.2. Thus the stalk $IC_{X_{\mathcal{O}},T_{\mathcal{O}}}|_q$ is a direct sum of shifted constant sheaves $C_{q,T_{\mathcal{O}}}$ (14.3(i')). Hence the corresponding statement is true on the induced space U, which proves the lemma.

15.16. Remark. Notice that $H_{T,c}(i_k^* F) = \oplus_{dim\mathcal{O}=k} H_{T,c}(F|_{\mathcal{O}})$. Moreover, it follows from the above lemma and from theorem 13.11.4 that

$$H_{T,c}(F|_{\mathcal{O}})[k] = A_\sigma \otimes_R IC(X)|_{\mathcal{O}},$$

where $k = dim\mathcal{O}$, $IC(X)|_{\mathcal{O}}$ is the stalk of $IC(X)$ at a point on $\mathcal{O}$ and σ is the cone in Φ corresponding to the orbit $\mathcal{O}$.

15.17. Corollary. *The A-module $H_{T,c}(i_k^* F)$ is pure of codimension k.*

15.18. Corollary. *The codimension of the support of the A-module $H_{T,c}(F_k)$ in $Spec A$ is $\leq k$. Hence*

$$Ext_A^i(H_{T,c}(F_k), A) = 0, \quad for \ i < k.$$

Proof of corollary. The A–module $H_{T,c}(F_k)$ may be computed using the spectral sequence associated to the filtration

$$U^n \subset \ldots \subset U^{k+1} \subset U^k.$$

The E_1 term consists of A-modules $H_{T,c}(i_s^* F), s \geq k$, which are pure of codimension s (lemma 15.17) and hence have support of codimension s. Hence the codimension of the support of $H_{T,c}(F_k)$ is $\geq k$ (in fact $= k$). This proves the corollary.

Now we can prove proposition 15.14. We use induction on k. For $k = 0$ we have the sequence

$$0 \to H_{T,c}(F) \to H_{T,c}(i_0^* F) \to H_{T,c}(F_1)[1] \to 0.$$

This is a piece of the long exact sequence of cohomology arising from the triangle (1) (for $k = 0$). Hence it suffices to prove that the connecting homomorphism $H_{T,c}(F_1)[1] \overset{\delta}{\to} H_{T,c}(F)[1]$ is zero. We know that $H_{T,c}(F)$ is a free A-module (remark 15.8(1)) and that the support of $H_{T,c}(F_1)$ has codimension 1 in $Spec A$. Hence there exists no nonzero map δ. This shows that the sequence above is short exact. By considering the long exact sequence of $\text{Ext}_A(\cdot, A)$ applied to this short sequence we find that $H_{T,c}(F_1)$ is pure of codimension 1 (use corollary 15.17). This finishes the proof for $k = 0$.

Suppose we proved the lemma for $k - 1$. Consider the sequence

$$0 \to H_{T,c}(F_k)[k] \to H_{T,c}(i_k^* F)[k] \to H_{T,c}(F_{k+1})[k + 1] \to 0$$

By induction we know that $H_{T,c}(F_k)$ is pure of codimension k. On the other hand $H_{T,c}(F_{k+1})$ has support of codimension $k+1$. Hence the connecting homomorphism $H_{T,c}(F_{k+1})[k + 1] \overset{\delta}{\to} H_{T,c}(F_k)[k + 1]$ is zero and the sequence is short exact. Applying $\text{Ext}_A(\cdot, A)$ to this sequence we find

$$0 \to \text{Ext}_A^k(H_{T,c}(i_k^* F)[k]) \to \text{Ext}_A^k(H_{T,c}(F_k)[k]) \to \text{Ext}_A^{k+1}(H_{T,c}(F_{k+1})[k + 1]) \to 0$$

so that $H_{T,c}(F_{k+1})$ is pure of codimension $k + 1$. This finishes the induction step and proves proposition 15.14 and theorem 15.11.

15.19. Definition. Let

$$0 \to \mathcal{L}^{-n} \overset{\partial^{-n}}{\to} \mathcal{L}^{-n+1} \overset{\partial^{-n+1}}{\to} \ldots \overset{\partial^{-1}}{\to} \mathcal{L}^0 \to 0$$

be the complex (3) shifted by n, i.e.

$$\mathcal{L}^{-k} := H_{T,c}(i_{n-k}^* F)[n - k]$$

$$\partial^{-k} := d^{n-k}$$

Fix a cone $\sigma \in \Phi$ of dimension k. Consider the restriction $F|_{\mathcal{O}_\sigma}$ of F to the corresponding orbit $\mathcal{O}_\sigma \subset X$. By remark 15.16 above we know that $H_{T,c}(F|_{\mathcal{O}_\sigma})[n - k]$ is a free A_σ module with the basis $IC(X)|_{\mathcal{O}_\sigma}$. We put $\mathcal{L}_\sigma := H_{T,c}(F|_{\mathcal{O}_\sigma})[n - k]$, so that

$$\mathcal{L}^{-k} = \oplus_{dim\sigma=k} \mathcal{L}_\sigma.$$

Clearly, the differential

$$\partial^{-k} : \mathcal{L}^{-k} \to \mathcal{L}^{-k+1}$$

maps

$$\partial^{-k} : \mathcal{L}_\sigma \to \oplus_{\tau \subset \sigma} \mathcal{L}_\tau.$$

In view of theorem 15.11 above the theorem 15.7 follows from the following

15.20. Proposition. *The complex $\mathcal{L}$ is minimal.*

Proof. We have to check conditions (a), (b) of definition 15.3.

(a) $\mathcal{L}^0 = H_{T,c}(i_n^* F)[n] = H_{T,c}(C_T[n])[n] = I\!\!R[n]$ (13.11.4).

(b) Let $\sigma \in \Phi$ be a cone of dimension k. Let

$$\mathcal{L}(\sigma) : \mathcal{L}_\sigma \xrightarrow{\partial_\sigma^{-k}} \underset{dim\tau=k-1}{\oplus_{\tau \subset \sigma}} \mathcal{L}_\tau \xrightarrow{\partial_\sigma^{-k+1}} \cdots \to \mathcal{L}_0 \to 0$$

be the restriction of the complex $\mathcal{L}$ to σ as on 15.2. Let $I_\sigma = Ker\partial_\sigma^{-k+1}$. We need to show that ∂_σ^{-k} induces an isomorphism.

(∗) $\qquad\qquad\qquad \partial_\sigma^{-k} : \mathcal{L}_\sigma/m\mathcal{L}_\sigma \xrightarrow{\sim} I_\sigma/mI_\sigma$

Put $\mathcal{O} = \mathcal{O}_\sigma$. Let $U \subset X$ be the open subset as in the proof of lemma 15.15. Geometrically the complex $\mathcal{L}(\sigma)$ corresponds to the "restriction" of the complex (3) to the open set U. Since

$$U = T \times_{T_\sigma} X_\mathcal{O}$$

(see the proof of lemma 15.15 above), the complex $\mathcal{L}(\sigma)$ is obtained from the corresponding complex (3) for $X_\mathcal{O}$ by restricting scalars from A_σ to A and shifting (13.11.4).

Hence we may (and will) assume that $\mathcal{O} = q$ is the T-fixed point (dim$\sigma = k = n$) in the affine toric variety U. By theorem 15.11 the complex $\mathcal{L}(\sigma)$ is exact except at $\mathcal{L}_\sigma$. Put $V = U - \{q\}$. The kernell I_σ is equal to $H_{T,c}(F|_V[1])$ and the map

$$\partial_\sigma^{-n} : \mathcal{L}_\sigma \to I_\sigma$$

is the canonical boundary map

$$H_{T,c}(F_q) \to H_{T,c}(F|_V)[1],$$

which induces an isomorphism module the maximal ideal (corollary 14.8). This proves proposition 15.20 and theorem 15.7.

Bibliography.

[BBD] A.Beilinson, J.Bernstein, P.Deligne. Faiceaux perverse, Asterisque, 1982.

[Bo1] A.Borel *et al.*. Intersection cohomology, Birkhauser, 1984.

[Bo2] A.Borel *et al.*. Algebraic D-modules, Academic Press, 1987.

[Bo3] A.Borel. Sur la cohomologie des espaces fibres principaux et des espaces homogenes de group de Lie compacts, Ann. Math., 57 (1953), 115-207.

[D1] P.Deligne. Theorie de Hodge III, Publ. Math. IHES. 44 (1974), 5-78.

[D2] P.Deligne. La conjecture de Weil II, Publ. Math. IHES. 52 (1980), 137-252.

[D3] P.Deligne. Cohomologie a supports propres, SGA4, LNM 305, 1973.

[Dan] V.Danilov. Geometry of toric varieties, Russian Math. Surveys (Uspehy), t.XXXIII, 2(200), 1978.

[DL] I.Dolgachev, V.Lunts. A character formula for the representation of the Weyl group in the cohomology of the associated toric variety, (to appear).

[DV] M.Duflo, M.Vergne. Sur le foncteur de Zuckerman, C.R.A.S. Paris, t.304, Serie I, n 16, 1987, 467-469.

[Gi] J.Giraud. Cohomologie non abelienne, Springer-Verlag, 1971.

[Go] R.Godement. Topologie algebrique et theorie des faisceaux, Hermann, Paris, 1958.

[GoMa] M.Goresky, R.MacPherson. Intersection Homology II, Invent. Math. 71, 77-129.

[Groth] A.Grothendieck. Sur quelques points d'algebre homologique, Tohoku Math. Journal, second series, vol.9, n 2, 3, 1957.

[H] R.Hartshorne. Residues and duality, LNM 20, 1966.

[Il] L.Illusie. Complexe cotangent et deformations, LNM 239 (1971), 283 (1972).

[KKMS-D] G.Kempf, F.Knudsen, D.Mumford, B.Saint-Donat. Toroidal embeddings, I, LNM 339 (1973).

[Lu] D.Luna. Slices etales, Bull. Soc. Math. France, 1973, memoire 33.

[S] E.Spanier. Algebraic topology, McGraw-Hill book company, 1966.

[Sp] N.Spaltenstein. Resolutions of unbounded complexes, Compositio Mathematica 65:121-154 (1988).

[Ve1] J.-L.Verdier. Categories derivees, LNM 569, 1977.

[Ve2] J.-L.Verdier. Dualite dans la cohomologie des espaces localement compact, Sem. Bourbaki, 1965/66, n 300.

Index